T0202893

Second Edition

SHALLOW FOUNDATIONS

Bearing Capacity and Settlement

Second Edition

SHALLOW FOUNDATIONS

Bearing Capacity and Settlement

Braja M. Das

CRC Press
Taylor & Francis Group
Boca Raton London New York

CRC Press is an imprint of the
Taylor & Francis Group, an **informa** business

CRC Press
Taylor & Francis Group
6000 Broken Sound Parkway NW, Suite 300
Boca Raton, FL 33487-2742

First issued in paperback 2019

ISBN-13: 978-1-4200-7006-4 (hbk)
ISBN-13: 978-0-367-38595-8 (pbk)

Library of Congress Cataloging-in-Publication Data

Das, Braja M., 1941-
    Shallow foundations bearing capacity and settlement / Braja M. Das. -- 2nd ed.
        p. cm.
    Includes bibliographical references and index.
    ISBN 978-1-4200-7006-4 (hardcover : alk. paper)
    1. Foundations. 2. Settlement of structures. 3. Soil mechanics. I. Title.

TA775.D2275 2009
624.1'5--dc22   2009000683

Visit the Taylor & Francis Web site at
http://www.taylorandfrancis.com

and the CRC Press Web site at
http://www.crcpress.com

Dedication

To our granddaughter, Elizabeth Madison

Contents

Preface

Shallow Foundations: Bearing Capacity and Settlement was originally published with a 1999 copyright and was intended for use as a reference book by university faculty members and graduate students in geotechnical engineering as well as by consulting engineers. During the last ten years, the text has served that constituency well. More recently there have been several requests to update the material and prepare a new edition. This edition of the text has been developed in response to those requests.

The text is divided into eight chapters. Chapters 2, 3, and 4 present various theories developed during the past 50 years for estimating the ultimate bearing capacity of shallow foundations under various types of loading and subsoil conditions. In this edition new details relating to the variation of the bearing capacity factor N_γ published more recently have been added and compared in Chapter 2. This chapter also has a broader overview and discussion on shape factors as well as scale effects on the bearing capacity tests conducted on granular soils. Ultimate bearing capacity relationships for shallow foundations subjected to eccentric and inclined loads have been added in Chapter 3. Published results of recent laboratory tests relating to the ultimate bearing capacity of square and circular foundations on granular soil of limited thickness underlain by a rigid rough base have been included in Chapter 4.

Chapter 5 discusses the principles for estimating the settlement of foundations—both elastic and consolidation. Westergaard's solution for stress distribution caused by a point load and uniformly loaded flexible circular and rectangular areas has been added. Procedures to estimate the elastic settlement of foundations on granular soil have been fully updated and presented in a rearranged form. These procedures include those based on the correlation with standard penetration resistance, strain influence factor, and the theory of elasticity.

Chapter 6 discusses dynamic bearing capacity and associated settlement. Also included in this chapter are some details regarding permanent foundation settlement due to cyclic and transient loadings derived from experimental observations obtained from laboratory and field tests.

During the past 25 years, steady progress has been made to evaluate the possibility of using reinforcement in granular soil to increase the ultimate and allowable bearing capacities of shallow foundations and also to reduce their settlement under various types of loading conditions. The reinforcement materials include galvanized steel strips and geogrids. Chapter 7 presents the state of the art on this subject.

Shallow foundations (such as transmission tower foundations) are on some occasions subjected to uplifting forces. The theories relating to the estimations of the ultimate uplift capacity of shallow foundations in granular and clay soils are presented in Chapter 8.

Example problems to illustrate the theories are given in each chapter.

I am grateful to my wife, Janice, for typing the manuscript and preparing the necessary artwork.

About the Author

Professor Braja M. Das received his Ph.D. in geotechnical engineering from the University of Wisconsin, Madison, USA. In 2006, after serving 12 years as dean of the College of Engineering and Computer Science at California State University, Sacramento, Professor Das retired and now lives in the Las Vegas, Nevada, area.

A fellow and life member in the American Society of Civil Engineers (ASCE), Professor Das served on the ASCE's Shallow Foundations Committee, Deep Foundations Committee, and Grouting Committee. He was also a member of the ASCE's editorial board for the *Journal of Geotechnical Engineering*. From 2000 to 2006, he was the coeditor of *Geotechnical and Geological Engineering—An International Journal* published by Springer in the Netherlands. Now an emeritus member of the Committee of Chemical and Mechanical Stabilization of the Transportation Research Board of the National Research Council of the United States, he served as committee chair from 1995 to 2001. He is also a life member of the American Society for Engineering Education. He was recently named the editor-in-chief of a new journal—the *International Journal of Geotechnical Engineering*—published by J. Ross Publishing of Florida (USA). The first issue of the journal was released in October 2007.

Dr. Das has received numerous awards for teaching excellence. He is the author of several geotechnical engineering text and reference books and has authored numerous technical papers in the area of geotechnical engineering. His primary areas of research include shallow foundations, earth anchors, and geosynthetics.

1 Introduction

1.1 SHALLOW FOUNDATIONS—GENERAL

The lowest part of a structure that transmits its weight to the underlying soil or rock is the foundation. Foundations can be classified into two major categories—*shallow foundations* and *deep foundations*. Individual footings (Figure 1.1), square or rectangular in plan, that support columns and strip footings that support walls and other similar structures are generally referred to as shallow foundations. *Mat foundations*, also considered shallow foundations, are reinforced concrete slabs of considerable structural rigidity that support a number of columns and wall loads. Several types of mat foundations are currently used. Some of the common types are shown schematically in Figure 1.2 and include

1. Flat plate (Figure 1.2a). The mat is of uniform thickness.
2. Flat plate thickened under columns (Figure 1.2b).
3. Beams and slab (Figure 1.2c). The beams run both ways, and the columns are located at the intersections of the beams.
4. Flat plates with pedestals (Figure 1.2d).
5. Slabs with basement walls as a part of the mat (Figure 1.2e). The walls act as stiffeners for the mat.

When the soil located immediately below a given structure is weak, the load of the structure may be transmitted to a greater depth by *piles* and *drilled shafts,* which are considered *deep foundations*. This book is a compilation of the theoretical and experimental evaluations presently available in the literature as they relate to the load-bearing capacity and settlement of shallow foundations.

The shallow foundation shown in Figure 1.1 has a width B and a length L. The depth of embedment below the ground surface is equal to D_f. Theoretically, when B/L is equal to zero (that is, $L = \infty$), a plane strain case will exist in the soil mass supporting the foundation. For most practical cases, when $B/L \leq 1/5$ to $1/6$, the plane strain theories will yield fairly good results. Terzaghi[1] defined a shallow foundation as one in which the depth D_f is less than or equal to the width B ($D_f/B \leq 1$). However, research studies conducted since then have shown that D_f/B can be as large as 3 to 4 for shallow foundations.

1.2 TYPES OF FAILURE IN SOIL AT ULTIMATE LOAD

Figure 1.3 shows a shallow foundation of width B located at a depth of D_f below the ground surface and supported by dense sand (or stiff, clayey soil). If this foundation is subjected to a load Q that is gradually increased, the load per unit area, $q = Q/A$

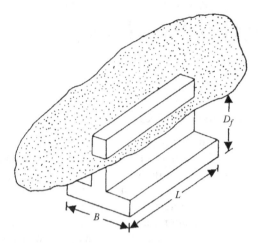

FIGURE 1.1 Individual footing.

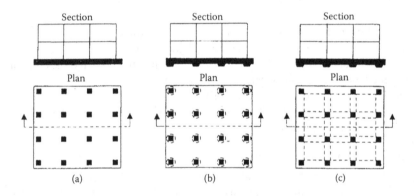

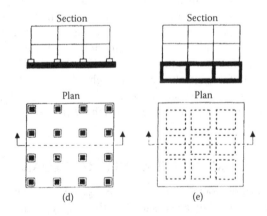

FIGURE 1.2 Various types of mat foundations: (a) flat plate; (b) flat plate thickened under columns; (c) beams and slab; (d) flat plate with pedestals; (e) slabs with basement walls.

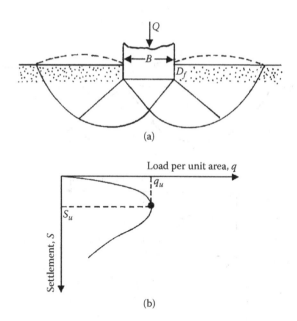

(a)

(b)

FIGURE 1.3 General shear failure in soil.

(A = area of the foundation), will increase and the foundation will undergo increased settlement. When q becomes equal to q_u at foundation settlement $S = S_u$, the soil supporting the foundation undergoes sudden shear failure. The failure surface in the soil is shown in Figure 1.3a, and the q versus S plot is shown in Figure 1.3b. This type of failure is called a *general shear failure*, and q_u is the *ultimate bearing capacity*. Note that, in this type of failure, a peak value of $q = q_u$ is clearly defined in the load-settlement curve.

If the foundation shown in Figure 1.3a is supported by a medium dense sand or clayey soil of medium consistency (Figure 1.4a), the plot of q versus S will be as shown in Figure 1.4b. Note that the magnitude of q increases with settlement up to $q = q'_u$, and this is usually referred to as the *first failure load*.[2] At this time, the developed failure surface in the soil will be as shown by the solid lines in Figure 1.4a. If the load on the foundation is further increased, the load-settlement curve becomes steeper and more erratic with the gradual outward and upward progress of the failure surface in the soil (shown by the jagged line in Figure 1.4b) under the foundation. When q becomes equal to q_u (ultimate bearing capacity), the failure surface reaches the ground surface. Beyond that, the plot of q versus S takes almost a linear shape, and a peak load is never observed. This type of bearing capacity failure is called a *local shear failure*.

Figure 1.5a shows the same foundation located on a loose sand or soft clayey soil. For this case, the load-settlement curve will be like that shown in Figure 1.5b. A peak value of load per unit area q is never observed. The ultimate bearing capacity q_u is

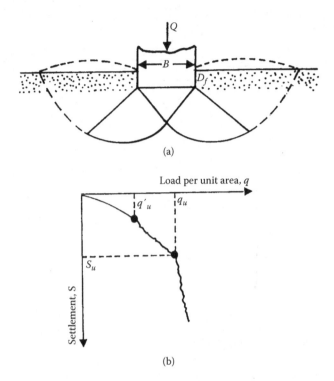

FIGURE 1.4 Local shear failure in soil.

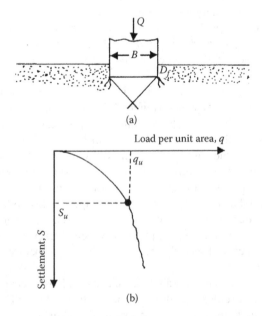

FIGURE 1.5 Punching shear failure in soil.

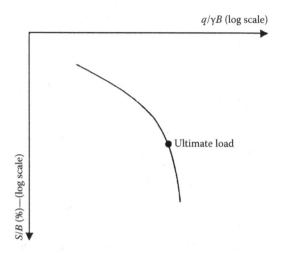

FIGURE 1.6 Nature of variation of $q/\gamma B$ with S/B in a log-log plot.

defined as the point where $\Delta S/\Delta q$ becomes the largest and remains almost constant thereafter. This type of failure in soil is called a *punching shear failure*. In this case the failure surface never extends up to the ground surface. In some cases of punching shear failure, it may be difficult to determine the ultimate load per unit area q_u from the q versus S plot shown in Figure 1.5. DeBeer[3] recommended a very consistent ultimate load criteria in which a plot of log $q/\gamma B$ versus log S/B is prepared (γ= unit weight of soil). The ultimate load is defined as the point of break in the log–log plot as shown in Figure 1.6.

The nature of failure in soil at ultimate load is a function of several factors such as the strength and the relative compressibility of the soil, the depth of the foundation (D_f) in relation to the foundation width B, and the width-to-length ratio (B/L) of the foundation. This was clearly explained by Vesic,[2] who conducted extensive laboratory model tests in sand. The summary of Vesic's findings is shown in a slightly different form in Figure 1.7. In this figure D_r is the relative density of sand, and the hydraulic radius R of the foundation is defined as

$$R = \frac{A}{P} \qquad (1.1)$$

where
 A = area of the foundation = BL
 P = perimeter of the foundation = $2(B + L)$

Thus,

$$R = \frac{BL}{2(B+L)} \qquad (1.2)$$

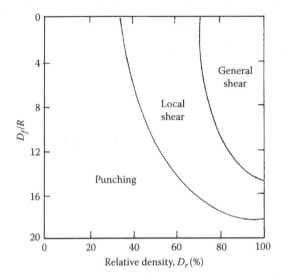

FIGURE 1.7 Nature of failure in soil with relative density of sand D_r and D_f/R.

for a square foundation $B = L$. So,

$$R = \frac{B}{4} \tag{1.3}$$

From Figure 1.7 it can be seen that when $D_f/R \geq$ about 18, punching shear failure occurs in all cases irrespective of the relative density of compaction of sand.

1.3 SETTLEMENT AT ULTIMATE LOAD

The settlement of the foundation at ultimate load S_u is quite variable and depends on several factors. A general sense can be derived from the laboratory model test results in *sand* for surface foundations ($D_f/B = 0$) provided by Vesic[4] and which are presented in Figure 1.8. From this figure it can be seen that, for any given foundation, a decrease in the relative density of sand results in an increase in the settlement at ultimate load. DeBeer[3] provided laboratory test results of circular surface foundations having diameters of 38 mm, 90 mm, and 150 mm on sand at various relative densities (D_r) of compaction. The results of these tests are summarized in Figure 1.9. It can be seen that, in general, for granular soils the settlement at ultimate load S_u increases with the increase in the width of the foundation B.

Based on laboratory and field test results, the approximate ranges of values of S_u in various types of soil are given in Table 1.1.

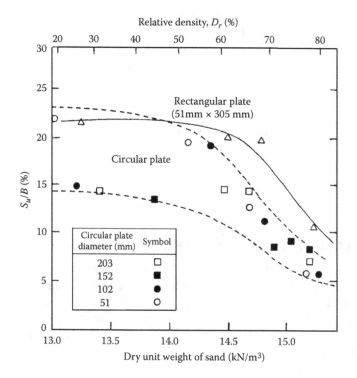

FIGURE 1.8 Variation of $\frac{S_u}{B}$ for surface foundation $\left(\frac{D_f}{B}=0\right)$ on sand. Source: From Vesic, A. S. 1973. Analysis of ultimate loads on shallow foundations. J. Soil Mech. Found. Div., ASCE, 99(1): 45.

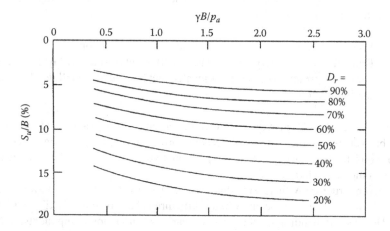

FIGURE 1.9 DeBeer's laboratory test results on circular surface foundations on sand—variation of $\frac{S_u}{B}$ with $\frac{\gamma B}{p_a}$ and D_r. Note: B = diameter of circular foundation; p_a = atmospheric pressure $\approx 100 \text{ kN/m}^2$; γ = unit weight of sand.

TABLE 1.1
Approximate Ranges of S_u

Soil	$\dfrac{D_f}{B}$	$\dfrac{S_u}{B}$ (%)
Sand	0	5–12
Sand	Large	25–28
Clay	0	4–8
Clay	Large	15–20

1.4 ULTIMATE AND ALLOWABLE BEARING CAPACITIES

For a given foundation to perform to its optimum capacity, one must ensure that the load per unit area of the foundation does not exceed a limiting value, thereby causing shear failure in soil. This limiting value is the ultimate bearing capacity q_u. Considering the ultimate bearing capacity and the uncertainties involved in evaluating the shear strength parameters of the soil, the allowable bearing capacity q_{all} can be obtained as

$$q_{all} = \frac{q_u}{FS} \tag{1.4}$$

A factor of safety of three to four is generally used. However, based on limiting settlement conditions, there are other factors that must be taken into account in deriving the allowable bearing capacity. The total settlement S_t of a foundation will be the sum of the following:

1. Elastic, or immediate, settlement S_e (described in section 1.3), and
2. Primary and secondary consolidation settlement S_c of a clay layer (located below the groundwater level) if located at a reasonably small depth below the foundation.

Most building codes provide an allowable settlement limit for a foundation, which may be well below the settlement derived corresponding to q_{all} given by equation (1.4). Thus, the bearing capacity corresponding to the allowable settlement must also be taken into consideration.

A given structure with several shallow foundations may undergo uniform settlement (Figure 1.10a). This occurs when a structure is built over a very rigid structural mat. However, depending on the loads on various foundation components, a structure may experience *differential settlement*. A foundation may undergo uniform tilt (Figure 1.10b) or nonuniform settlement (Figure 1.10c). In these cases, the angular

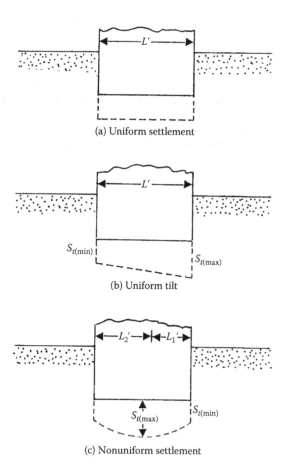

(a) Uniform settlement

(b) Uniform tilt

(c) Nonuniform settlement

FIGURE 1.10 Settlements of a structure.

distortion Δ can be defined as

$$\Delta = \frac{S_{t(max)} - S_{t(min)}}{L'} \quad \text{(for uniform tilt)} \tag{1.5}$$

and

$$\Delta = \frac{S_{t(max)} - S_{t(min)}}{L_1'} \quad \text{(for nonuniform tilt)} \tag{1.6}$$

Limits for allowable differential settlements of various structures are also available in building codes. Thus, the final decision on the allowable bearing capacity of a foundation will depend on (a) the ultimate bearing capacity, (b) the allowable settlement, and (c) the allowable differential settlement for the structure.

REFERENCES

1. Terzaghi, K. 1943. *Theoretical Soil Mechanics*. New York: Wiley.
2. Vesic, A. S. 1973. Analysis of ultimate loads on shallow foundations. *J. Soil Mech. Found. Div.,* ASCE, 99(1): 45.
3. DeBeer, E. E. 1967. Proefondervindelijke bijdrage tot de studie van het gransdraagvermogen van zand onder funderingen op staal, Bepaling von der vormfactor s_b. *Annales des Travaux Publics de Belgique* 6: 481.
4. Vesic, A. S. 1963. Bearing capacity of deep foundations in sand. *Highway Res. Rec.,* National Research Council, Washington, D.C. 39:12.

2 Ultimate Bearing Capacity Theories — Centric Vertical Loading

2.1 INTRODUCTION

Over the last 60 years, several bearing capacity theories for estimating the ultimate bearing capacity of shallow foundations have been proposed. This chapter summarizes some of the important works developed so far. The cases considered in this chapter assume that the soil supporting the foundation extends to a great depth and also that the foundation is subjected to centric vertical loading. The variation of the ultimate bearing capacity in anisotropic soils is also considered.

2.2 TERZAGHI'S BEARING CAPACITY THEORY

In 1948 Terzaghi[1] proposed a well-conceived theory to determine the ultimate bearing capacity of a shallow, rough, rigid, *continuous* (strip) foundation supported by a homogeneous soil layer extending to a great depth. Terzaghi defined a shallow foundation as a foundation where the width B is equal to or less than its depth D_f. The failure surface in soil at ultimate load (that is, q_u per unit area of the foundation) assumed by Terzaghi is shown in Figure 2.1. Referring to Figure 2.1, the failure area in the soil under the foundation can be divided into three major zones:

1. Zone *abc*. This is a triangular *elastic* zone located immediately below the bottom of the foundation. The inclination of sides *ac* and *bc* of the wedge with the horizontal is $\alpha = \phi$ (soil friction angle).
2. Zone *bcf*. This zone is the Prandtl's radial shear zone.
3. Zone *bfg*. This zone is the *Rankine passive zone*. The *slip lines* in this zone make angles of $\pm (45 - \phi/2)$ with the horizontal.

Note that a Prandtl's radial shear zone and a Rankine passive zone are also located to the left of the elastic triangular zone *abc*; however, they are not shown in Figure 2.1.

Line *cf* is an arc of a *log spiral* and is defined by the equation

$$r = r_0 e^{\theta \tan \phi} \tag{2.1}$$

Lines *bf* and *fg* are straight lines. Line *fg* actually extends up to the ground surface. Terzaghi assumed that the soil located above the bottom of the foundation could be replaced by a surcharge $q = \gamma D_f$.

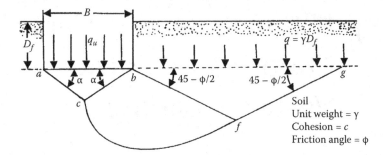

FIGURE 2.1 Failure surface in soil at ultimate load for a continuous rough rigid foundation as assumed by Terzaghi.

The shear strength of the soil can be given as

$$s = \sigma' \tan \phi + c \qquad (2.2)$$

where
  σ' = effective normal stress
  c = cohesion

The ultimate bearing capacity q_u of the foundation can be determined if we consider faces ac and bc of the triangular wedge abc and obtain the passive force on each face required to cause failure. Note that the passive force P_p will be a function of the surcharge $q = \gamma D_f$, cohesion c, unit weight γ, and angle of friction of the soil ϕ. So, referring to Figure 2.2, the passive force P_p on the face bc per unit length of the foundation at a right angle to the cross section is

$$P_p = P_{pq} + P_{pc} + P_{p\gamma} \qquad (2.3)$$

where
  P_{pq}, P_{pc}, and $P_{p\gamma}$ = passive force contributions of q, c, and γ, respectively

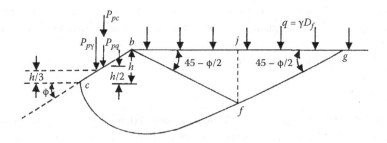

FIGURE 2.2 Passive force on the face bc of wedge abc shown in Figure 2.1.

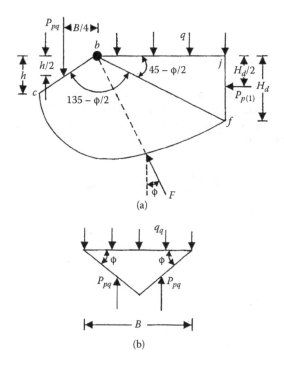

FIGURE 2.3 Determination of P_{pq} ($\phi \neq 0$, $\gamma = 0$, $q \neq 0$, $c = 0$).

It is important to note that the directions of P_{pq}, P_{pc}, and $P_{p\gamma}$ are vertical since the face bc makes an angle ϕ with the horizontal, and P_{pq}, P_{pc}, and $P_{p\gamma}$ must make an angle ϕ to the normal drawn to bc. In order to obtain P_{pq}, P_{pc}, and $P_{p\gamma}$, the method of superposition can be used; however, it will not be an exact solution.

2.2.1   RELATIONSHIP FOR P_{pq} ($\phi \neq 0$, $\gamma = 0$, $q \neq 0$, $c = 0$)

Consider the free body diagram of the soil wedge $bcfj$ shown in Figure 2.2 (also shown in Figure 2.3). For this case, the center of the log spiral (of which cf is an arc) will be at point b. The forces *per unit length of the wedge bcfj* due to the surcharge q only are shown in Figure 2.3a, and they are

1. P_{pq}
2. Surcharge q
3. The Rankine passive force $P_{p(1)}$
4. The frictional resisting force F along the arc cf

The Rankine passive force $P_{p(1)}$ can be expressed as

$$P_{p(1)} = qK_p H_d = qH_d \tan^2\left(45 + \frac{\phi}{2}\right) \qquad (2.4)$$

where

$$H_d = \overline{fj}$$

K_p = Rankine passive earth pressure coefficient = $\tan^2(45 + \phi/2)$

According to the property of a log spiral defined by the equation $r = r_0 e^{\theta \tan \phi}$, the radial line at any point makes an angle ϕ with the normal; hence, the line of action of the frictional force F will pass through b (the center of the log spiral as shown in Figure 2.3a). Taking the moment of all forces about point b:

$$P_{pq}\left(\frac{B}{4}\right) = q(\overline{bj})\left(\frac{\overline{bj}}{2}\right) + P_{p(1)}\frac{H_d}{2} \tag{2.5}$$

let

$$\overline{bc} = r_0 = \left(\frac{B}{2}\right)\sec \phi \tag{2.6}$$

From equation (2.1):

$$\overline{bf} = r_1 = r_0 e^{\left(\frac{3\pi}{4} - \frac{\phi}{2}\right)\tan \phi} \tag{2.7}$$

So,

$$\overline{bj} = r_1 \cos\left(45 - \frac{\phi}{2}\right) \tag{2.8}$$

and

$$H_d = r_1 \sin\left(45 - \frac{\phi}{2}\right) \tag{2.9}$$

Combining equations (2.4), (2.5), (2.8), and (2.9):

$$\frac{P_{pq}B}{4} = \frac{qr_1^2 \cos^2\left(45 - \frac{\phi}{2}\right)}{2} + \frac{qr_1^2 \sin^2\left(45 - \frac{\phi}{2}\right)\tan^2\left(45 + \frac{\phi}{2}\right)}{2}$$

or

$$P_{pq} = \frac{4}{B}\left[qr_1^2 \cos^2\left(45 - \frac{\phi}{2}\right)\right] \tag{2.10}$$

Now, combining equations (2.6), (2.7), and (2.10):

$$P_{pq} = qB \sec^2 \phi\left[e^{2\left(\frac{3\pi}{4} - \frac{\phi}{2}\right)\tan \phi}\right]\left[\cos^2\left(45 - \frac{\phi}{2}\right)\right] = \frac{qBe^{2\left(\frac{3\pi}{4} - \frac{\phi}{2}\right)\tan \phi}}{4\cos^2\left(45 + \frac{\phi}{2}\right)} \tag{2.11}$$

Considering the stability of the elastic wedge abc under the foundation as shown in Figure 2.3b

$$q_q(B \times 1) = 2P_{pq}$$

where

q_q = load per unit area on the foundation, or

$$q_q = \frac{2P_{pq}}{B} = q \underbrace{\left[\frac{e^{2\left(\frac{3\pi}{4} - \frac{\phi}{2}\right)\tan\phi}}{2\cos^2\left(45 + \frac{\phi}{2}\right)} \right]}_{N_q} = qN_q \tag{2.12}$$

2.2.2 RELATIONSHIP FOR P_{pc} ($\phi \neq 0$, $\gamma = 0$, $q = 0$, $c \neq 0$)

Figure 2.4 shows the free body diagram for the wedge $bcfj$ (also refer to Figure 2.2). As in the case of P_{pq}, the center of the arc of the log spiral will be located at point b. The forces on the wedge, which are due to cohesion c, are also shown in Figure 2.4, and they are

1. Passive force P_{pc}
2. Cohesive force $\overline{C} = c(\overline{bc} \times 1)$

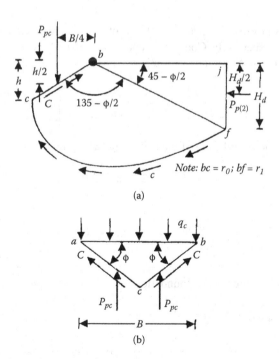

(a)

(b)

FIGURE 2.4 Determination of P_{pc} ($\phi \neq 0$, $\gamma = 0$, $q = 0$, $c \neq 0$).

3. Rankine passive force due to cohesion

$$P_{p(2)} = 2c\sqrt{K_p}H_d = 2cH_d \tan\left(45+\frac{\phi}{2}\right)$$

4. Cohesive force per unit area c along arc cf

Taking the moment of all the forces about point b:

$$P_{pc}\left(\frac{B}{4}\right) = P_{p(2)}\left[\frac{r_1 \sin\left(45-\dfrac{\phi}{2}\right)}{2}\right] + M_c \qquad (2.13)$$

where

$$M_c = \text{moment due to cohesion } c \text{ along arc } cf = \frac{c}{2\tan\phi}\left(r_1^2 - r_0^2\right) \qquad (2.14)$$

So,

$$P_{pc}\left(\frac{B}{4}\right) = \left[2cH_d \tan\left(45+\frac{\phi}{2}\right)\right]\left[\frac{r_1 \sin\left(45-\dfrac{\phi}{2}\right)}{2}\right] + \left(\frac{c}{2\tan\phi}\right)\left(r_1^2 - r_0^2\right) \qquad (2.15)$$

The relationships for H_d, r_0, and r_1 in terms of B and ϕ are given in equations (2.9), (2.6), and (2.7), respectively. Combining equations (2.6), (2.7), (2.9), and (2.15), and noting that $\sin^2(45-\phi/2) \times \tan(45+\phi/2) = \frac{1}{2}\cos\phi$,

$$P_{pc} = Bc(\sec^2\phi)\left[e^{2\left(\frac{3\pi}{4}-\frac{\phi}{2}\right)\tan\phi}\right]\left(\frac{\cos\phi}{2}\right) + \left(\frac{Bc}{2\tan\phi}\right)\sec^2\phi\left[e^{2\left(\frac{3\pi}{4}-\frac{\phi}{2}\right)\tan\phi}\right] \qquad (2.16)$$

Considering the equilibrium of the soil wedge abc (Figure 2.4b):

$$q_c(B\times1) = 2C\sin\phi + 2P_{pc}$$

or

$$q_cB = cB\sec\phi\sin\phi + 2P_{pc} \qquad (2.17)$$

where
q_c = load per unit area of the foundation

Combining equations (2.16) and (2.17):

$$q_c = c\sec\phi\, e^{2\left(\frac{3\pi}{4}-\frac{\phi}{2}\right)\tan\phi} + \frac{c\sec^2\phi}{\tan\phi}e^{2\left(\frac{3\pi}{4}-\frac{\phi}{2}\right)\tan\phi} - \frac{c\sec^2\phi}{\tan\phi} + c\tan\phi \qquad (2.18)$$

or

$$q_c = ce^{2\left(\frac{3\pi}{4}-\frac{\phi}{2}\right)\tan\phi}\left[\sec\phi+\frac{\sec^2\phi}{\tan\phi}\right]-c\left[\frac{\sec^2\phi}{\tan\phi}-\tan\phi\right] \quad (2.19)$$

However,

$$\sec\phi+\frac{\sec^2\phi}{\tan\phi}=\frac{1}{\cos\phi}+\frac{1}{\cos\phi\sin\phi}=\cot\phi\left(\frac{1+\sin\phi}{\cos^2\phi}\right)=\cot\phi\left[\frac{1}{2\cos^2\left(45+\frac{\phi}{2}\right)}\right] \quad (2.20)$$

Also,

$$\frac{\sec^2\phi}{\tan\phi}-\tan\phi=\cot\phi(\sec^2\phi-\tan^2\phi)=\cot\phi\left(\frac{1}{\cos^2\phi}-\frac{\sin^2\phi}{\cos^2\phi}\right)$$

$$=\cot\phi\left(\frac{\cos^2\phi}{\cos^2\phi}\right)=\cot\phi \quad (2.21)$$

Substituting equations (2.20) and (2.21) into equation (2.19)

$$q_c=c\cot\phi\left[\frac{e^{2\left(\frac{3\pi}{4}-\frac{\phi}{2}\right)\tan\phi}}{2\cos^2\left(45+\frac{\phi}{2}\right)}-1\right]=cN_c=c\cot\phi(N_q-1) \quad (2.22)$$

2.2.3 RELATIONSHIP FOR $P_{p\gamma}$ ($\phi \neq 0$, $\gamma \neq 0$, $q = 0$, $c = 0$)

Figure 2.5a shows the free body diagram of wedge *bcfj*. Unlike the free body diagrams shown in Figures 2.3 and 2.4, the center of the log spiral of which *bf* is an arc is at a point *O* along line *bf* and not at *b*. This is because the minimum value of $P_{p\gamma}$ has to be determined by several trials. Point *O* is only one trial center. The forces per unit length of the wedge that need to be considered are

1. Passive force $P_{p\gamma}$
2. The weight *W* of wedge *bcfj*
3. The resultant of the frictional resisting force *F* acting along arc *cf*
4. The Rankine passive force $P_{p(3)}$

The Rankine passive force $P_{p(3)}$ can be given by the relation

$$P_{p(3)}=\frac{1}{2}\gamma H_d^2\tan^2\left(45+\frac{\phi}{2}\right) \quad (2.23)$$

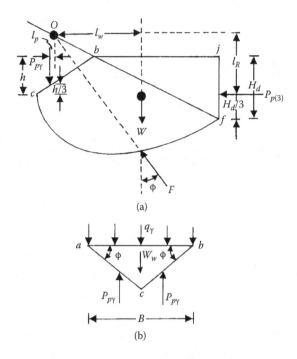

FIGURE 2.5 Determination of $P_{p\gamma}$ ($\phi \neq 0$, $\gamma \neq 0$, $q = 0$, $c = 0$).

Also note that the line of action of force F will pass through O. Taking the moment of all forces about O:

$$P_{p\gamma}l_p = Wl_w + P_{p(3)}l_R$$

or

$$P_{p\gamma} = \frac{1}{l_p}[Wl_w + P_{p(3)}l_R] \tag{2.24}$$

If a number of trials of this type are made by changing the location of the center of the log spiral O along line bf, then the minimum value of $P_{p\gamma}$ can be determined.

Considering the stability of wedge abc as shown in Figure 2.5, we can write that

$$q_\gamma B = 2P_{p\gamma} - W_w \tag{2.25}$$

where

q_γ = force per unit area of the foundation
W_w = weight of wedge abc

However,

$$W_w = \frac{B^2}{4} \gamma \tan \phi \qquad (2.26)$$

So,

$$q_\gamma = \frac{1}{B} \left(2 P_{p\gamma} - \frac{B^2}{4} \gamma \tan \phi \right) \qquad (2.27)$$

The passive force $P_{p\gamma}$ can be expressed in the form

$$P_{p\gamma} = \frac{1}{2} \gamma h^2 K_{p\gamma} = \frac{1}{2} \gamma \left(\frac{B \tan \phi}{2} \right)^2 K_{p\gamma} = \frac{1}{8} \gamma B^2 K_{p\gamma} \tan^2 \phi \qquad (2.28)$$

where $K_{p\gamma}$ = passive earth pressure coefficient

Substituting equation (2.28) into equation (2.27)

$$q_\gamma = \frac{1}{B} \left(\frac{1}{4} \gamma B^2 K_{p\gamma} \tan^2 \phi - \frac{B^2}{4} \gamma \tan \phi \right) = \frac{1}{2} \gamma B \left(\frac{1}{2} K_{p\gamma} \tan^2 \phi - \frac{\tan \phi}{2} \right) = \frac{1}{2} \gamma B N_\gamma$$

$$(2.29)$$

2.2.4 Ultimate Bearing Capacity

The ultimate load per unit area of the foundation (that is, the ultimate bearing capacity q_u) for a soil with cohesion, friction, and weight can now be given as

$$q_u = q_q + q_c + q_\gamma \qquad (2.30)$$

Substituting the relationships for q_q, q_c, and q_γ given by equations (2.12), (2.22), and (2.29) into equation (2.30) yields

$$q_u = c N_c + q N_q + \frac{1}{2} \gamma B N_\gamma \qquad (2.31)$$

where

N_c, N_q, and N_γ = bearing capacity factors, and

$$N_q = \frac{e^{2\left(\frac{3\pi}{4} - \frac{\phi}{2} \right) \tan \phi}}{2 \cos^2 \left(45 + \frac{\phi}{2} \right)} \qquad (2.32)$$

$$N_c = \cot \phi (N_q - 1) \qquad (2.33)$$

$$N_\gamma = \frac{1}{2} K_{p\gamma} \tan^2 \phi - \frac{\tan \phi}{2} \qquad (2.34)$$

Table 2.1 gives the variations of the bearing capacity factors with soil friction angle ϕ given by equations (2.32), (2.33), and (2.34). The values of N_γ were obtained by Kumbhojkar.[2]

Krizek[3] gave simple empirical relations for Terzaghi's bearing capacity factors N_c, N_q, and N_γ with a maximum deviation of 15%. They are as follows:

$$N_c = \frac{228 + 4.3\phi}{40 - \phi} \qquad (2.35a)$$

$$N_q = \frac{40 + 5\phi}{40 - \phi} \qquad (2.35b)$$

$$N_\gamma = \frac{6\phi}{40 - \phi} \qquad (2.35c)$$

where
ϕ = soil friction angle, in degrees

Equations (2.35a), (2.35b), and (2.35c) are valid for $\phi = 0$ to $35°$. Thus, substituting equation (2.35) into (2.31),

$$q_u = \frac{(228 + 4.3\phi)c + (40 + 5\phi)q + 3\phi\gamma B}{40 - \phi} \quad \text{(for } \phi = 0° \text{ to } 35°) \qquad (2.36)$$

For foundations that are rectangular or circular in plan, a plane strain condition in soil at ultimate load does not exist. Therefore, Terzaghi[1] proposed the following relationships for square and circular foundations:

$$q_u = 1.3cN_c + qN_q + 0.4\gamma BN_\gamma \quad \text{(square foundation; plan } B \times B) \qquad (2.37)$$

and

$$q_u = 1.3cN_c + qN_q + 0.3\gamma BN_\gamma \quad \text{(circular foundation; diameter } B) \qquad (2.38)$$

Since Terzaghi's founding work, numerous experimental studies to estimate the ultimate bearing capacity of shallow foundations have been conducted. Based on these studies, it appears that Terzaghi's assumption of the failure surface in soil at ultimate load is essentially correct. However, the angle α that sides ac and bc of the wedge (Figure 2.1) make with the horizontal is closer to $45 + \phi/2$ and not ϕ, as assumed by Terzaghi. In that case, the nature of the soil failure surface would be as shown in Figure 2.6.

The method of superposition was used to obtain the bearing capacity factors N_c, N_q, and N_γ. For derivations of N_c and N_q, the center of the arc of the log spiral cf is located at the edge of the foundation. That is not the case for the derivation of N_γ. In effect, two different surfaces are used in deriving equation (2.31); however, it is on the safe side.

TABLE 2.1
Terzaghi's Bearing Capacity Factors—Equations
(2.32), (2.33), and (2.34)

ϕ	N_c	N_q	N_γ
0	5.70	1.00	0.00
1	6.00	1.10	0.01
2	6.30	1.22	0.04
3	6.62	1.35	0.06
4	6.97	1.49	0.10
5	7.34	1.64	0.14
6	7.73	1.81	0.20
7	8.15	2.00	0.27
8	8.60	2.21	0.35
9	9.09	2.44	0.44
10	9.61	2.69	0.56
11	10.16	2.98	0.69
12	10.76	3.29	0.85
13	11.41	3.63	1.04
14	12.11	4.02	1.26
15	12.86	4.45	1.52
16	13.68	4.92	1.82
17	14.60	5.45	2.18
18	15.12	6.04	2.59
19	16.57	6.70	3.07
20	17.69	7.44	3.64
21	18.92	8.26	4.31
22	20.27	9.19	5.09
23	21.75	10.23	6.00
24	23.36	11.40	7.08
25	25.13	12.72	8.34
26	27.09	14.21	9.84
27	29.24	15.90	11.60
28	31.61	17.81	13.70
29	34.24	19.98	16.18
30	37.16	22.46	19.13
31	40.41	25.28	22.65
32	44.04	28.52	26.87
33	48.09	32.23	31.94
34	52.64	36.50	38.04
35	57.75	41.44	45.41
36	63.53	47.16	54.36
37	70.01	53.80	65.27
38	77.50	61.55	78.61
39	85.97	70.61	95.03
40	95.66	81.27	115.31
41	106.81	93.85	140.51
42	119.67	108.75	171.99
43	134.58	126.50	211.56
44	151.95	147.74	261.60
45	172.28	173.28	325.34
46	196.22	204.19	407.11
47	224.55	241.80	512.84
48	258.28	287.85	650.87
49	298.71	344.63	831.99
50	347.50	415.14	1072.80

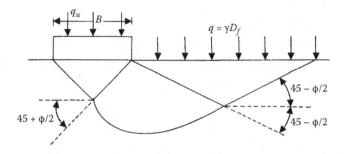

FIGURE 2.6 Modified failure surface in soil supporting a shallow foundation at ultimate load.

2.3 TERZAGHI'S BEARING CAPACITY THEORY FOR LOCAL SHEAR FAILURE

It is obvious from section 2.2 that Terzaghi's bearing capacity theory was obtained assuming general shear failure in soil. However, Terzaghi[1] suggested the following relationships for local shear failure in soil:

Strip foundation ($B/L = 0$; $L = $ length of foundation):

$$q_u = c'N_c' + qN_q' + \frac{1}{2}\gamma BN_\gamma' \qquad (2.39)$$

Square foundation ($B = L$):

$$q_u = 1.3c'N_c' + qN_q' + 0.4\gamma BN_\gamma' \qquad (2.40)$$

Circular foundation ($B = $ diameter):

$$q_u = 1.3c'N_c' + qN_q' + 0.3\gamma BN_\gamma' \qquad (2.41)$$

where

N_c', N_q', and N_γ' = modified bearing capacity factors

$c' = 2c/3$

The modified bearing capacity factors can be obtained by substituting $\phi' = \tan^{-1}(0.67 \tan \phi)$ for ϕ in equations (2.32), (2.33), and (2.34). The variations of N_c', N_q', and N_γ' with ϕ are shown in Table 2.2.

Vesic[4] suggested a better mode to obtain ϕ' for estimating N_c' and N_q' for foundations on sand in the forms

$$\phi' = \tan^{-1}(k \tan \phi) \qquad (2.42)$$

$$k = 0.67 + D_r - 0.75D_r^2 \quad \text{(for } 0 \leq D_r \leq 0.67) \qquad (2.43)$$

where

$D_r = $ relative density

TABLE 2.2

Terzaghi's Modified Bearing Capacity Factors N'_c, N'_q, and N'_γ

ϕ	N'_c	N'_q	N'_γ
0	5.70	1.00	0.00
1	5.90	1.07	0.005
2	6.10	1.14	0.02
3	6.30	1.22	0.04
4	6.51	1.30	0.055
5	6.74	1.39	0.074
6	6.97	1.49	0.10
7	7.22	1.59	0.128
8	7.47	1.70	0.16
9	7.74	1.82	0.20
10	8.02	1.94	0.24
11	8.32	2.08	0.30
12	8.63	2.22	0.35
13	8.96	2.38	0.42
14	9.31	2.55	0.48
15	9.67	2.73	0.57
16	10.06	2.92	0.67
17	10.47	3.13	0.76
18	10.90	3.36	0.88
19	11.36	3.61	1.03
20	11.85	3.88	1.12
21	12.37	4.17	1.35
22	12.92	4.48	1.55
23	13.51	4.82	1.74
24	14.14	5.20	1.97
25	14.80	5.60	2.25
26	15.53	6.05	2.59
27	16.03	6.54	2.88
28	17.13	7.07	3.29
29	18.03	7.66	3.76
30	18.99	8.31	4.39
31	20.03	9.03	4.83
32	21.16	9.82	5.51
33	22.39	10.69	6.32
34	23.72	11.67	7.22
35	25.18	12.75	8.35
36	26.77	13.97	9.41
37	28.51	15.32	10.90
38	30.43	16.85	12.75
39	32.53	18.56	14.71
40	34.87	20.50	17.22
41	37.45	22.70	19.75
42	40.33	25.21	22.50
43	43.54	28.06	26.25
44	47.13	31.34	30.40
45	51.17	35.11	36.00
46	55.73	39.48	41.70
47	60.91	44.54	49.30
48	66.80	50.46	59.25
49	73.55	57.41	71.45
50	81.31	65.60	85.75

2.4   MEYERHOF'S BEARING CAPACITY THEORY

In 1951, Meyerhof published a bearing capacity theory that could be applied to rough, shallow, and deep foundations. The failure surface at ultimate load under a continuous shallow foundation assumed by Meyerhof[5] is shown in Figure 2.7. In this figure abc is the elastic triangular wedge shown in Figure 2.6, bcd is the radial shear zone with cd being an arc of a log spiral, and bde is a mixed shear zone in which the shear varies between the limits of radial and plane shears depending on the depth and roughness of the foundation. The plane be is called an *equivalent free surface.* The normal and shear stresses on plane be are p_o and s_o, respectively. The superposition method is used to determine the contribution of cohesion c, p_o, γ, and ϕ on the ultimate bearing capacity q_u of the *continuous* foundation and is expressed as

$$q_u = cN_c + qN_q + \tfrac{1}{2}\gamma BN_\gamma \tag{2.44}$$

where
$\quad N_c$, N_q, and N_γ = bearing capacity factors
$\quad\quad B$ = width of the foundation

2.4.1   DERIVATION OF N_c AND N_q ($\phi \neq 0$, $\gamma = 0$, $p_o \neq 0$, $c \neq 0$)

For this case, the center of the log spiral arc [equation (2.1)] is taken at b. Also, it is assumed that along be

$$s_o = m(c + p_o \tan \phi) \tag{2.45}$$

where
$\quad c$ = cohesion
$\quad \phi$ = soil friction angle
$\quad m$ = degree of mobilization of shear strength ($0 \leq m \leq 1$)

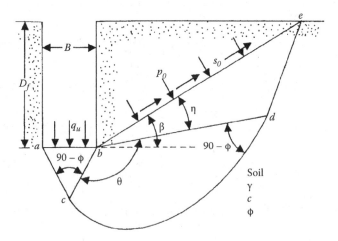

FIGURE 2.7 Slip line fields for a rough continuous foundation.

Now consider the linear zone *bde* (Figure 2.8a). Plastic equilibrium requires that the shear strength s_1 under the normal stress p_1 is fully mobilized, or

$$s_1 = c + p_1 \tan \phi \qquad (2.46)$$

Figure 2.8b shows the Mohr's circle representing the stress conditions on zone *bde*. Note that P is the pole. The traces of planes *bd* and *be* are also shown in the figure. For the Mohr's circle,

$$R = \frac{s_1}{\cos \phi} \qquad (2.47)$$

where
R = radius of the Mohr's circle

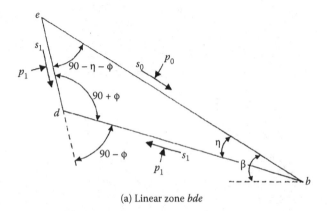

(a) Linear zone *bde*

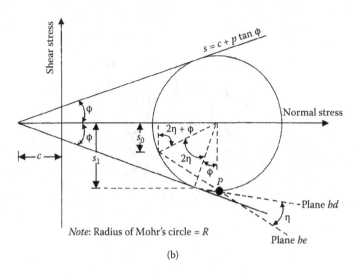

Note: Radius of Mohr's circle = R

(b)

FIGURE 2.8 Determination of N_q and N_c.

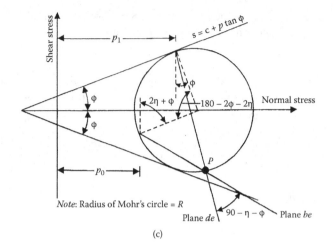

(c)

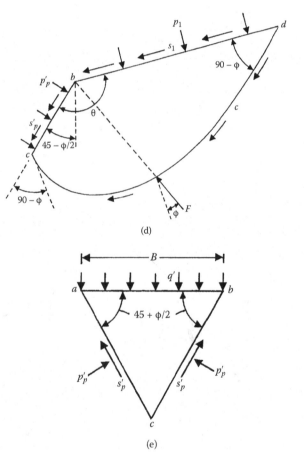

(d)

(e)

FIGURE 2.8 (Continued).

Also,

$$s_o = R\cos(2\eta + \phi) = \frac{s_1 \cos(2\eta + \phi)}{\cos\phi} \tag{2.48}$$

Combining equations (2.45), (2.46), and (2.48):

$$\cos(2\eta + \phi) = \frac{s_o \cos\phi}{c + p_1 \tan\phi} = \frac{m(c + p_o \tan\phi)\cos\phi}{c + p_1 \tan\phi} \tag{2.49}$$

Again, referring to the trace of plane de (Figure 2.8c),

$$s_1 = R\cos\phi$$

$$R = \frac{c + p_1 \tan\phi}{\cos\phi} \tag{2.50}$$

Note that

$$p_1 + R\sin\phi = p_0 + R\sin(2\eta + \phi)$$

$$p_1 = R[\sin(2\eta + \phi) - \sin\phi] + p_o = \frac{c + p_1 \tan\phi}{\cos\phi}[\sin(2\eta + \phi) - \sin\phi] + p_o \tag{2.51}$$

Figure 2.8d shows the free body diagram of zone bcd. Note that the normal and shear stresses on the face bc are P_p' and s_p', or

$$s_p' = c + p_p' \tan\phi$$

or

$$p_p' = (s_p' - c)\cot\phi \tag{2.52}$$

Taking the moment of all forces about b,

$$P_1\left(\frac{r_1^2}{2}\right) - p_p'\left(\frac{r_0^2}{2}\right) + M_c = 0 \tag{2.53}$$

where

$$r_0 = \overline{bc}$$

$$r_1 = \overline{bd} = r_0 e^{\theta \tan\phi} \tag{2.54}$$

It can be shown that

$$M_c = \frac{c}{2\tan\phi}(r_1^2 - r_0^2) \tag{2.55}$$

Substituting equations (2.54) and (2.55) into equation (2.53) yields

$$p'_p = p_1 e^{2\theta \tan \phi} + c \cot \phi (e^{2\theta \tan \phi} - 1) \tag{2.56}$$

Combining equations (2.52) and (2.56)

$$s'_p = (c + p_1 \tan \phi) e^{2\theta \tan \phi} \tag{2.57}$$

Figure 2.8e shows the free body diagram of wedge abc. Resolving the forces in the vertical direction,

$$2 p'_p \left[\frac{\frac{B}{2}}{\cos\left(45 + \frac{\phi}{2}\right)} \right] \cos\left(45 + \frac{\phi}{2}\right) + 2 s'_p \left[\frac{\frac{B}{2}}{\cos\left(45 + \frac{\phi}{2}\right)} \right] \sin\left(45 + \frac{\phi}{2}\right) = q'B$$

where
$q' = $ load per unit area of the foundation, or

$$q' = p'_p + s'_p \cot\left(45 - \frac{\phi}{2}\right) \tag{2.58}$$

Substituting equations (2.51), (2.52), and (2.57) into equation (2.58) and further simplifying yields

$$q' = c \left\{ \cot \phi \underbrace{\left[\frac{(1 + \sin \phi) e^{2\theta \tan \phi}}{1 - \sin \phi \sin(2\eta + \phi)} - 1 \right]}_{N_c} \right\} + p_o \underbrace{\left[\frac{(1 + \sin \phi) e^{2\theta \tan \phi}}{1 - \sin \phi \sin(2\eta + \phi)} \right]}_{N_q} = cN_c + p_o N_q$$

$$\tag{2.59}$$

where
$N_c, N_q = $ bearing capacity factors

The bearing capacity factors will depend on the degree of mobilization m of shear strength on the equivalent free surface. This is because m controls η. From equation (2.49)

$$\cos(2\eta + \phi) = \frac{m(c + p_o \tan \phi) \cos \phi}{c + p_1 \tan \phi}$$

For $m = 0$, $\cos(2\eta + \phi) = 0$, or

$$\eta = 45 - \frac{\phi}{2} \tag{2.60}$$

For $m = 1$, $\cos(2\eta + \phi) = \cos \phi$, or

$$\eta = 0 \qquad (2.61)$$

Also, the factors N_c and N_q are influenced by the angle of inclination of the equivalent free surface β. From the geometry of Figure 2.7,

$$\theta = 135° + \beta - \eta - \frac{\phi}{2} \qquad (2.62)$$

From equation (2.60), for $m = 0$, the value of η is $(45 - \phi/2)$. So,

$$\theta = 90° + \beta \quad \text{(for } m = 0) \qquad (2.63)$$

Similarly, for $m = 1$ [since $\eta = 0$; equation (2.61)]:

$$\theta = 135° + \beta - \frac{\phi}{2} \quad \text{(for } m = 1) \qquad (2.64)$$

Figures 2.9 and 2.10 show the variations of N_c and N_q with ϕ, β, and m. It is of interest to note that, if we consider the surface foundation condition (as done in Figures 2.3 and 2.4 for Terzaghi's bearing capacity equation derivation), then $\beta = 0$ and $m = 0$. So, from equation (2.63),

$$\theta = \frac{\pi}{2} \qquad (2.65)$$

Hence, for $m = 0$, $\eta = 45 - \phi/2$, and $\theta = \pi/2$, the expressions for N_c and N_q are as follows (surface foundation condition):

$$N_q = e^{\pi \tan \phi} \left(\frac{1 + \sin \phi}{1 - \sin \phi} \right) \qquad (2.66)$$

and

$$N_c = (N_q - 1) \cot \phi \qquad (2.67)$$

Equations (2.66) and (2.67) are the same as those derived by Reissner[6] for N_q and Prandtl[7] for N_c. For this condition $p_o = \gamma D_f = q$. So equation (2.59) becomes

$$q' = \underbrace{cN_c}_{\text{Eq. (2.66)}} + \underbrace{qN_q}_{\text{Eq. (2.67)}} \qquad (2.68)$$

2.4.2   DERIVATION OF N_γ ($\phi \neq 0$, $\gamma \neq 0$, $p_o = 0$, $c = 0$)

N_γ is determined by trial and error as in the case of the derivation of Terzaghi's bearing capacity factor N_γ (section 2.2). Referring to Figure 2.11a, following is a

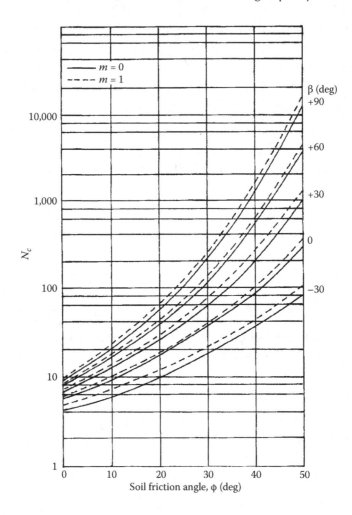

FIGURE 2.9 Meyerhof's bearing capacity factor—variation of N_c with β, ϕ, and m [equation (2.59)].

step-by-step approach for the derivation of N_γ:

1. Choose values for ϕ and the angle β (such as +30°, +40°, −30°…).
2. Choose a value for m (such as $m = 0$ or $m = 1$).
3. Determine the value of θ from equation (2.63) or (2.64) for $m = 0$ or $m = 1$, as the case may be.
4. With known values of θ and β, draw lines bd and be.
5. Select a trial center such as O and draw an arc of a log spiral connecting points c and d. The log spiral follows the equation $r = r_0 e^{\theta \tan \phi}$.
6. Draw line de. Note that lines bd and de make angles of $90 - \phi$ due to the restrictions on slip lines in the linear zone bde. Hence the trial failure surface is not, in general, continuous at d.

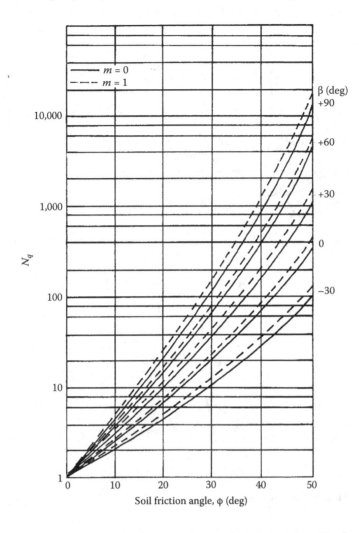

FIGURE 2.10 Meyerhof's bearing capacity factor—variation of N_q with β, ϕ, and m [equation (2.59)].

7. Consider the trial wedge $bcdf$. Determine the following forces per unit length of the wedge at right angles to the cross section shown: (a) weight of wedge $bcdf$—W, and (b) Rankine passive force on the face df—$P_{p(R)}$.
8. Take the moment of the forces about the trial center of the log spiral O, or

$$P_{p\gamma} = \frac{Wl_w + P_{p(R)}l_R}{l_p} \qquad (2.69)$$

where

$P_{p\gamma}$ = passive force due to γ and ϕ only

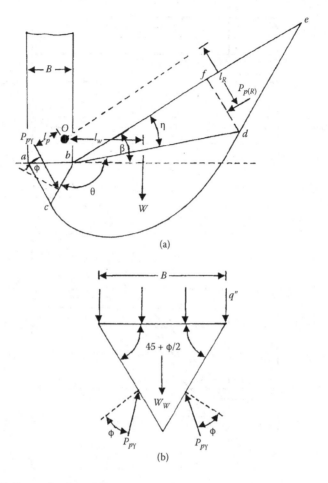

FIGURE 2.11 Determination of N_γ.

Note that <u>the</u> line of action of $P_{p\gamma}$ acting on the face bc is located at a distance of $2bc/3$.

9. For given values of β, ϕ, and m, and by changing the location of point O
(that is, the center of the log spiral), repeat steps 5 through 8 to obtain the
minimum value of $P_{p\gamma}$.

Refer to Figure 2.11b. Resolve the forces acting on the triangular wedge abc in
the vertical direction, or

$$q'' = \frac{\gamma B}{2}\left[\frac{4P_{p\gamma}\sin\left(45+\frac{\phi}{2}\right)}{\gamma B^2} - \frac{1}{2}\tan\left(45+\frac{\phi}{2}\right)\right] = \frac{1}{2}\gamma BN_\gamma \qquad (2.70)$$

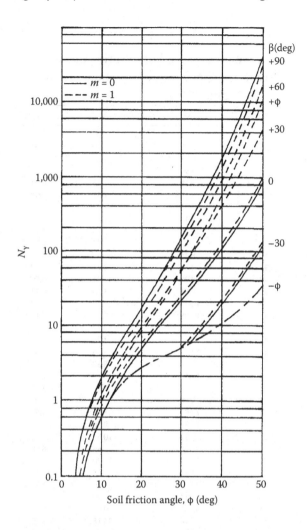

FIGURE 2.12 Meyerhof's bearing capacity factor—variation of N_γ with β, ϕ, and m [equation (2.70)].

where
    q'' = force per unit area of the foundation
    N_γ = bearing capacity factor

Note that W_w is the weight of wedge abc in Figure 2.11b. The variation of N_γ (determined in the above manner) with β, ϕ, and m is given in Figure 2.12.

Combining equations (2.59) and (2.70), the ultimate bearing capacity of a continuous foundation (for the condition $c \neq 0$, $\gamma \neq 0$, and $\phi \neq 0$) can be given as

$$q_u = q' + q'' = cN_c + p_oN_q + \tfrac{1}{2}\gamma BN_\gamma$$

TABLE 2.3
Variation of Meyerhof's Bearing Capacity
Factors N_c, N_q, and N_γ [Equations (2.66),
(2.67), and (2.72)]

ϕ	N_c	N_q	N_γ
0	5.14	1.00	0.00
1	5.38	1.09	0.002
2	5.63	1.20	0.01
3	5.90	1.31	0.02
4	6.19	1.43	0.04
5	6.49	1.57	0.07
6	6.81	1.72	0.11
7	7.16	1.88	0.15
8	7.53	2.06	0.21
9	7.92	2.25	0.28
10	8.35	2.47	0.37
11	8.80	2.71	0.47
12	9.28	2.97	0.60
13	9.81	3.26	0.74
14	10.37	3.59	0.92
15	10.98	3.94	1.13
16	11.63	4.34	1.38
17	12.34	4.77	1.66
18	13.10	5.26	2.00
19	13.93	5.80	2.40
20	14.83	6.40	2.87
21	15.82	7.07	3.42
22	16.88	7.82	4.07
23	18.05	8.66	4.82
24	19.32	9.60	5.72
25	20.72	10.66	6.77
26	22.25	11.85	8.00
27	23.94	13.20	9.46
28	25.80	14.72	11.19
29	27.86	16.44	13.24
30	30.14	18.40	15.67
31	32.67	20.63	18.56
32	35.49	23.18	22.02
33	38.64	26.09	26.17
34	42.16	29.44	31.15
35	46.12	33.30	37.15
36	50.59	37.75	44.43
37	55.63	42.92	53.27
38	61.35	48.93	64.07
39	67.87	55.96	77.33
40	75.31	64.20	93.69
41	83.86	73.90	113.99
42	93.71	85.38	139.32
43	105.11	99.02	171.14
44	118.37	115.31	211.41
45	133.88	134.88	262.74
46	152.10	158.51	328.73
47	173.64	187.21	414.32
48	199.26	222.31	526.44
49	229.93	265.51	674.91
50	266.89	319.07	873.84

The above equation is the same form as equation (2.44). Similarly, for *surface foundation conditions* (that is, $\beta = 0$ and $m = 0$), the ultimate bearing capacity of a continuous foundation can be given as

$$q_u = \underset{\text{Eq. (2.68)}}{q'} + \underset{\text{Eq. (2.70)}}{q''} = \underset{\text{Eq. (2.67)}}{cN_c} + \underset{\text{Eq. (2.66)}}{qN_q} + \tfrac{1}{2}\gamma BN_\gamma \qquad (2.71)$$

For shallow foundation designs, the ultimate bearing capacity relationship given by equation (2.71) is presently used. The variation of N_γ for surface foundation conditions (that is, $\beta = 0$ and $m = 0$) is given in Figure 2.12. In 1963 Meyerhof[8] suggested that N_γ could be approximated as

$$N_\gamma = (\underset{\text{Eq. (2.66)}}{N_q} - 1)\tan(1.4\phi) \qquad (2.72)$$

Table 2.3 gives the variations of N_c and N_q obtained from equations (2.66) and (2.67) and N_γ obtained from equation (2.72).

2.5   GENERAL DISCUSSION ON THE RELATIONSHIPS OF BEARING CAPACITY FACTORS

At this time, the general trend among geotechnical engineers is to accept the method of superposition as a suitable means to estimate the ultimate bearing capacity of shallow rough foundations. For *rough continuous foundations*, the nature of the failure surface in soil shown in Figure 2.6 has also found acceptance, as have Reissner's[6] and Prandtl's[7] solutions for N_c and N_q, which are the same as Meyerhof's[5] solution for surface foundations, or,

$$N_q = e^{\pi \tan\phi}\left(\frac{1 + \sin\phi}{1 - \sin\phi}\right) \qquad (2.66)$$

and

$$N_c = (N_q - 1)\cot\phi \qquad (2.67)$$

There has been considerable controversy over the theoretical values of N_γ. Hansen[9] proposed an approximate relationship for N_γ in the form

$$N_\gamma = 1.5N_c\tan^2\phi \qquad (2.73)$$

In the preceding equation, the relationship for N_c is that given by Prandtl's solution [equation (2.67)]. Caquot and Kerisel[10] assumed that the elastic triangular soil wedge under a rough continuous foundation is of the shape shown in Figure 2.6. Using integration of Boussinesq's differential equation, they presented numerical values of N_γ for various soil friction angles ϕ. Vesic[4] approximated their solutions in the form

$$N_\gamma = 2(N_q + 1)\tan\phi \qquad (2.74)$$

where
N_q is given by equation (2.66)

Equation (2.74) has an error not exceeding 5% for $20° < \phi < 40°$ compared to the exact solution. Lundgren and Mortensen[11] developed numerical methods (using the theory of plasticity) for the exact determination of rupture lines as well as the bearing capacity factor (N_γ) for particular cases. Chen[12] also gave a solution for N_γ in which he used the upper bound limit analysis theorem suggested by Drucker and Prager.[13] Biarez et al.[14] also recommended the following relationship for N_γ:

$$N_\gamma = 1.8(N_q - 1)\tan\phi \tag{2.75}$$

Booker[15] used the slip line method and provided numerical values of N_γ. Poulos et al.[16] suggested the following expression that approximates the numerical results of Booker[15]:

$$N_\gamma \approx 0.1045e^{9.6\phi} \tag{2.76}$$

where
ϕ is in radians
$N_\gamma = 0$ for $\phi = 0$

Recently Kumar[17] proposed another slip line solution based on Lundgren and Mortensen's failure mechanism.[11] Michalowski[18] also used the upper bound limit analysis theorem to obtain the variation of N_γ. His solution can be approximated as

$$N_\gamma = e^{(0.66 + 5.1\tan\phi)}\tan\phi \tag{2.77}$$

Hjiaj et al.[19] obtained a numerical analysis solution for N_γ. This solution can be approximated as

$$N_\gamma = e^{\frac{1}{6}(\pi + 3\pi^2\tan\phi)}(\tan\phi)^{\frac{2\pi}{5}} \tag{2.78}$$

Martin[20] used the method of characteristics to obtain the variations of N_γ. Salgado[21] approximated these variations in the form

$$N_\gamma = (N_q - 1)\tan(1.32\phi) \tag{2.79}$$

Table 2.4 gives a comparison of the N_γ values recommended by Meyerhof,[8] Terzaghi,[1] Vesic,[4] and Hansen.[9] Table 2.5 compares the variations of N_γ obtained by Chen,[12] Booker,[15] Kumar,[17] Michalowski,[18] Hjiaj et al.,[19] and Martin.[20]

The primary reason several theories for N_γ were developed, and their lack of correlation with experimental values, lies in the difficulty of selecting a representative value of the soil friction angle ϕ for computing bearing capacity. The parameter ϕ depends on many factors, such as intermediate principal stress condition, friction angle anisotropy, and curvature of the Mohr-Coulomb failure envelope.

It has been suggested that the plane strain soil friction angle ϕ_p, instead of ϕ_t, be used to estimate bearing capacity.[9] To that effect Vesic[4] raised the issue that this type of assumption might help explain the differences between the theoretical and experimental results for long rectangular foundations; however, it does not help to interpret results

TABLE 2.4
Comparison of N_γ Values (Rough Foundation)

Soil Friction Angle ϕ (deg)	N_γ			
	Terzaghi [Equation (2.34)]	Meyerhof [Equation (2.72)]	Vesic [Equation (2.74)]	Hansen [Equation (2.73)]
0	0.00	0.00	0.00	0.00
1	0.01	0.002	0.07	0.00
2	0.04	0.01	0.15	0.01
3	0.06	0.02	0.24	0.02
4	0.10	0.04	0.34	0.05
5	0.14	0.07	0.45	0.07
6	0.20	0.11	0.57	0.11
7	0.27	0.15	0.71	0.16
8	0.35	0.21	0.86	0.22
9	0.44	0.28	1.03	0.30
10	0.56	0.37	1.22	0.39
11	0.69	0.47	1.44	0.50
12	0.85	0.60	1.69	0.63
13	1.04	0.74	1.97	0.78
14	1.26	0.92	2.29	0.97
15	1.52	1.13	2.65	1.18
16	1.82	1.38	3.06	1.43
17	2.18	1.66	3.53	1.73
18	2.59	2.00	4.07	2.08
19	3.07	2.40	4.68	2.48
20	3.64	2.87	5.39	2.95
21	4.31	3.42	6.20	3.50
22	5.09	4.07	7.13	4.13
23	6.00	4.82	8.20	4.88
24	7.08	5.72	9.44	5.75
25	8.34	6.77	10.88	6.76
26	9.84	8.00	12.54	7.94
27	11.60	9.46	14.47	9.32
28	13.70	11.19	16.72	10.94
29	16.18	13.24	19.34	12.84
30	19.13	15.67	22.40	15.07
31	22.65	18.56	25.99	17.69
32	26.87	22.02	30.22	20.79
33	31.94	26.17	35.19	24.44
34	38.04	31.15	41.06	28.77
35	45.41	37.15	48.03	33.92
36	54.36	44.43	56.31	40.05
37	65.27	53.27	66.19	47.38
38	78.61	64.07	78.03	56.17
39	95.03	77.33	92.25	66.75
40	115.31	93.69	109.41	79.54
41	140.51	113.99	130.22	95.05
42	171.99	139.32	155.55	113.95
43	211.56	171.14	186.54	137.10
44	261.60	211.41	224.64	165.58
45	325.34	262.74	271.76	200.81

TABLE 2.5
Other N_γ Values (Rough Foundation)

SoilFriction Angle ϕ (deg)	Chen[12]	Booker[15]	Kumar[17]	Michalowski[18]	Hjiaj et al.[19]	Martin[20]
5	0.38	0.24	0.23	0.18	0.18	0.113
10	1.16	0.56	0.69	0.71	0.45	0.433
15	2.30	1.30	1.60	1.94	1.21	1.18
20	5.20	3.00	3.43	4.47	2.89	2.84
25	11.40	6.95	7.18	9.77	6.59	6.49
30	25.00	16.06	15.57	21.39	14.90	14.75
35	57.00	37.13	35.16	48.68	34.80	34.48
40	141.00	85.81	85.73	118.83	85.86	85.47
45	374.00	198.31	232.84	322.84	232.91	234.21

of tests with square or circular foundations. Ko and Davidson[22] also concluded that, when plane strain angles of internal friction are used in commonly accepted bearing capacity formulas, the bearing capacity for rough footings could be seriously overestimated for dense sands. To avoid the controversy Meyerhof[8] suggested the following:

$$\phi = \left[1.1 - 0.1 \left(\frac{B}{L} \right) \right] \phi_t$$

where
   ϕ_t = triaxial friction angle

2.6   OTHER BEARING CAPACITY THEORIES

Hu[23] proposed a theory according to which the base angle α of the triangular wedge below a rough foundation (refer to Figure 2.1) is a function of several parameters, or

$$\alpha = f(\gamma, \phi, q) \tag{2.80}$$

The minimum and maximum values of α can be given as follows:

$$\phi < \alpha_{min} < 45 + \frac{\phi}{2}$$

and

$$\alpha_{max} = 45 + \frac{\phi}{2}$$

The values of N_c, N_q, and N_γ determined by this procedure are shown in Figure 2.13.
   Balla[24] proposed a bearing capacity theory that was developed for an assumed failure surface in soil (Figure 2.14). For this failure surface, the curve cd was assumed

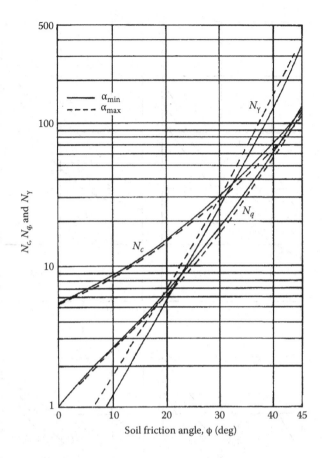

FIGURE 2.13 Hu's bearing capacity factors.

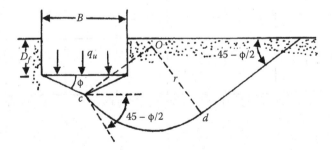

FIGURE 2.14 Nature of failure surface considered for Balla's bearing capacity theory.

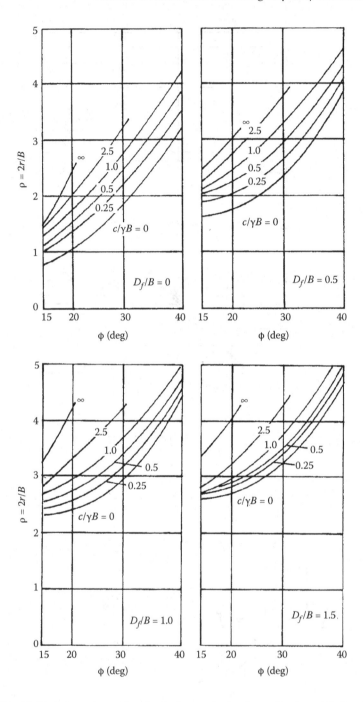

FIGURE 2.15 Variation of ρ with soil friction angle for determination of Balla's bearing capacity factors.

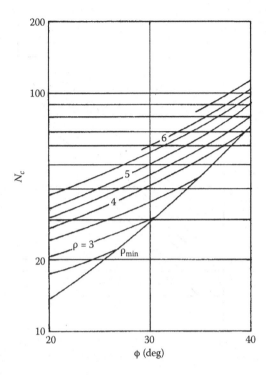

FIGURE 2.16 Balla's bearing capacity factor N_c.

to be an arc of a circle having a radius r. The bearing capacity solution was obtained using Kötter's equation to determine the distribution of the normal and tangential stresses on the slip surface. According to this solution for a continuous foundation,

$$q_u = cN_c + qN_q + \frac{1}{2}\gamma BN_\gamma$$

The bearing capacity factors can be determined as follows:

1. Obtain the magnitude of $c/B\gamma$ and D_f/B.
2. With the values obtained in step 1, go to Figure 2.15 to obtain the magnitude of $\rho = 2r/B$.
3. With known values of ρ, go to Figures 2.16, 2.17, and 2.18, respectively, to determine N_c, N_q, and N_γ.

2.7   SCALE EFFECTS ON ULTIMATE BEARING CAPACITY

The problem in estimating the ultimate bearing capacity becomes complicated if the *scale effect* is taken into consideration. Figure 2.19 shows the average variation of $N_\gamma/2$ with soil friction angle obtained from small footing tests in sand conducted

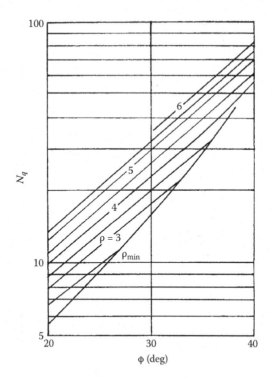

FIGURE 2.17 Balla's bearing capacity factor N_q.

in the laboratory at Ghent as reported by DeBeer.[25] For these tests, the values of ϕ were obtained from triaxial tests. This figure also shows the variation of $N_\gamma/2$ with ϕ obtained from tests conducted in Berlin and reported by Muhs[26] with footings having an area of 1 m². The soil friction angles for these tests were obtained from direct shear tests. It is interesting to note that:

1. For loose sand, the field test results of N_γ are higher than those obtained from small footing tests in the laboratory.
2. For dense sand, the laboratory tests provide higher values of N_γ compared to those obtained from the field.

The reason for the above observations can partially be explained by the fact that, in the field, *progressive rupture* in the soil takes place during the loading process. For loose sand at failure, the soil friction angle is higher than at the beginning of loading due to compaction. The reverse is true in the case of dense sand.

Figure 2.20 shows a comparison of several bearing capacity test results in sand compiled by DeBeer,[25] which are plots of N_γ with γB. For any given soil, the magnitude of N_γ decreases with B and remains constant for larger values of B.

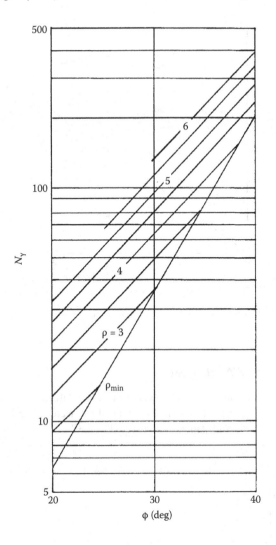

FIGURE 2.18 Balla's bearing capacity factor N_γ.

The reduction in N_γ for larger foundations may ultimately result in a substantial decrease in the ultimate bearing capacity that can primarily be attributed to the following reasons:

1. For larger-sized foundations, the rupture along the slip lines in soil is progressive, and the average shear strength mobilized (and thus ϕ) along a slip line decreases with the increase in B.
2. There are zones of weakness that exist in the soil under the foundation.
3. The curvature of the Mohr-Coulomb envelope.

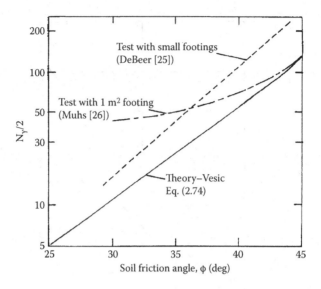

FIGURE 2.19 Comparison of N_γ obtained from tests with small footings and large footings (area = 1 m²) on sand.

2.8   EFFECT OF WATER TABLE

The preceding sections assume that the water table is located below the failure surface in the soil supporting the foundation. However, if the water table is present near the foundation, the terms q and γ in equations (2.31), (2.37), (2.38), (2.39) to (2.41), and

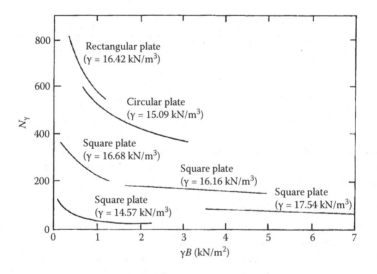

FIGURE 2.20 DeBeer's study on the variation of N_γ with γB.

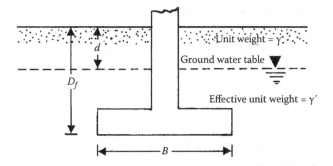

FIGURE 2.21 Effect of ground water table on ultimate bearing capacity.

(2.71) need to be modified. This process can be explained by referring to Figure 2.21, in which the water table is located at a depth d below the ground surface.

CASE I: $d = 0$

For $d = 0$, the term $q = \gamma D_f$ associated with N_q should be changed to $q = \gamma' D_f$ (γ' = effective unit weight of soil). Also, the term γ associated with N_γ should be changed to γ'.

CASE II: $0 < d \le D_f$

For this case, q will be equal to $\gamma d + (D_f - d)\, \gamma'$, and the term γ associated with N_γ should be changed to γ'.

CASE III: $D_f \le d \le D_f + B$

This condition is one in which the groundwater table is located at or below the bottom of the foundation. In such case, $q = \gamma D_f$ and the last term γ should be replaced by an average effective unit weight of soil $\bar{\gamma}$, or

$$\bar{\gamma} = \gamma' + \left(\frac{d - D_f}{B}\right)(\gamma - \gamma') \tag{2.81}$$

CASE IV: $d > D_f + B$

For $d > D_f + B$, $q = \gamma D_f$ and the last term should remain γ. This implies that the groundwater table has no effect on the ultimate capacity.

2.9   GENERAL BEARING CAPACITY EQUATION

The relationships to estimate the ultimate bearing capacity presented in the preceding sections are for continuous (strip) foundations. They do not give (a) the relationships for the ultimate bearing capacity for rectangular foundations (that is, $B/L > 0$; B = width and L = length), and (b) the effect of the depth of the foundation on the increase in the ultimate bearing capacity. Therefore, a general bearing capacity may be written as

$$q_u = cN_c\lambda_{cs}\lambda_{cd} + qN_q\lambda_{qs}\lambda_{qd} + \tfrac{1}{2}\gamma BN_\gamma\lambda_{\gamma s}\lambda_{\gamma d} \qquad (2.82)$$

where

$\lambda_{cs}, \lambda_{qs}, \lambda_{\gamma s}$ = shape factors
$\lambda_{cd}, \lambda_{qd}, \lambda_{\gamma d}$ = depth factors

Most of the shape and depth factors available in the literature are empirical and/or semi-empirical, and they are given in Table 2.6.

If equations (2.67), (2.66), and (2.74) are used for N_c, N_q, and N_γ, respectively, it is recommended that DeBeer's shape factors and Hansen's depth factors be used. However, if equations (2.67), (2.66), and (2.72) are used for bearing capacity factors N_c, N_q, and N_γ, respectively, then Meyerhof's shape and depth factors should be used.

EXAMPLE 2.1

A shallow foundation is 0.6 m wide and 1.2 m long. Given: $D_f = 0.6$ m. The soil supporting the foundation has the following parameters: $\phi = 25°$, $c = 48$ kN/m², and $\gamma = 18$ kN/m³. Determine the ultimate vertical load that the foundation can carry by using

   a. Prandtl's value of N_c [equation (2.67)], Reissner's value of N_q [equation (2.66)], Vesic's value of N_γ [equation (2.74)], and the shape and depth factors proposed by DeBeer and Hansen, respectively (Table 2.6)
   b. Meyerhof's values of N_c, N_q, and N_γ [equations (2.67), (2.66), and (2.72)] and the shape and depth factors proposed by Meyerhof[8] given in Table 2.6

Solution

From equation (2.82),

$$q_u = cN_c\lambda_{cs}\lambda_{cd} + qN_q\lambda_{qs}\lambda_{qd} + \tfrac{1}{2}\gamma BN_\gamma\lambda_{\gamma s}\lambda_{\gamma d}$$

Part a:
From Table 2.3 for $\phi = 25°$, $N_c = 20.72$ and $N_q = 10.66$. Also, from Table 2.4 for $\phi = 25°$, Vesic's value of $N_\gamma = 10.88$. DeBeer's shape factors are as follows:

$$\lambda_{cs} = 1 + \left(\frac{N_q}{N_c}\right)\left(\frac{B}{L}\right) = 1 + \left(\frac{10.66}{20.72}\right)\left(\frac{0.6}{1.2}\right) = 1.257$$

$$\lambda_{qs} = 1 + \left(\frac{B}{L}\right)\tan\phi = 1 + \left(\frac{0.6}{1.2}\right)\tan 25 = 1.233$$

$$\lambda_{\gamma s} = 1 - 0.4\left(\frac{B}{L}\right) = 1 - (0.4)\left(\frac{0.6}{1.2}\right) = 0.8$$

TABLE 2.6
Summary of Shape and Depth Factors

Factor	Relationship	Reference
Shape	For $\phi = 0°$: $\lambda_{cs} = 1 + 0.2\left(\dfrac{B}{L}\right)$ $\lambda_{qs} = 1$ $\lambda_{\gamma s} = 1$ For $\phi \geq 10°$: $\lambda_{cs} = 1 + 0.2\left(\dfrac{B}{L}\right)\tan^2\left(45 + \dfrac{\phi}{2}\right)$ $\lambda_{qs} = \lambda_{\gamma s} = 1 + 0.1\left(\dfrac{B}{L}\right)\tan^2\left(45 + \dfrac{\phi}{2}\right)$	Meyerhof[8]
	$\lambda_{cs} = 1 + \left(\dfrac{N_q}{N_c}\right)\left(\dfrac{B}{L}\right)$ [Note: Use equation (2.67) for N_c and equation (2.66) for N_q as given in Table 2.3] $\lambda_{qs} = 1 + \left(\dfrac{B}{L}\right)\tan\phi$ $\lambda_{\gamma s} = 1 - 0.4\left(\dfrac{B}{L}\right)$	DeBeer[27]
	$\lambda_{cs} = 1 + (1.8\tan^2\phi + 0.1)\left(\dfrac{B}{L}\right)^{0.5}$ $\lambda_{qs} = 1 + 1.9\tan^2\phi\left(\dfrac{B}{L}\right)^{0.5}$ $\lambda_{\gamma s} = 1 + (0.6\tan^2\phi - 0.25)\left(\dfrac{B}{L}\right)$ (for $\phi \leq 30°$) $\lambda_{\gamma s} = 1 + (1.3\tan^2\phi - 0.5)\left(\dfrac{L}{B}\right)^{1.5} e^{-\left(\frac{L}{B}\right)}$ (for $\phi > 30°$)	Michalowski[28]
	$\lambda_{cs} = 1 + C_1\left(\dfrac{B}{L}\right) + C_2\left(\dfrac{D_f}{B}\right)^{0.5}$ (for $\phi = 0$)	Salgado et al.[29]

(Continued)

TABLE 2.6 (CONTINUED)
Summary of Shape and Depth Factors

Factor	Relationship	Reference

Salgado et al.[29]

$\dfrac{B}{L}$	C_1	C_2
Circle	0.163	0.210
1.00	0.125	0.219
0.50	0.156	0.173
0.33	0.159	0.137
0.25	0.172	0.110
0.20	0.190	0.090

Depth   Meyerhof[8]

$$\text{For } \phi = 0^\circ: \quad \lambda_{cd} = 1 + 0.2\left(\frac{D_f}{B}\right)$$

$$\lambda_{qd} = \lambda_{\gamma d} = 1$$

$$\text{For } \phi \geq 10^\circ: \quad \lambda_{cd} = 1 + 0.2\left(\frac{D_f}{B}\right)\tan\left(45 + \frac{\phi}{2}\right)$$

$$\lambda_{qd} = \lambda_{\gamma d} = 1 + 0.1\left(\frac{D_f}{B}\right)\tan\left(45 + \frac{\phi}{2}\right)$$

$$\text{For } D_f/B \leq 1: \quad \lambda_{cd} = 1 + 0.4\left(\frac{D_f}{B}\right) \quad (\text{for } \phi = 0)$$

Hansen[9]

$$\lambda_{cd} = \lambda_{qd} - \frac{1 - \lambda_{qd}}{N_q \tan\phi}$$

$$\lambda_{qd} = 1 + 2\tan\phi(1 - \sin\phi)^2\left(\frac{D_f}{B}\right)$$

$$\lambda_{\gamma d} = 1$$

TABLE 2.6 (CONTINUED)
Summary of Shape and Depth Factors

Factor	Relationship	Reference
	For $D_f/B > 1$: $\lambda_{cd} = 1 + 0.4\tan^{-1}\left(\dfrac{D_f}{B}\right)$	Hansen[9]
	$\lambda_{qd} = 1 + 2\tan\phi(1-\sin\phi)^2\tan^{-1}\left(\dfrac{D_f}{B}\right)$	
	$\lambda_{\gamma d} = 1$	
	$\left[\text{Note: }\tan^{-1}\left(\dfrac{D_f}{B}\right)\text{ is in radians.}\right]$	
	$\lambda_{cd} = 1 + 0.27\left(\dfrac{D_f}{B}\right)^{0.5}$	Salgado et al.[29]

Hansen's depth factors are as follows:

$$\lambda_{qd} = 1 + 2\tan\phi(1-\sin\phi)^2\left(\frac{D_f}{B}\right) = 1 + 2(\tan 25)(1-\sin 25)^2\left(\frac{0.6}{0.6}\right) = 1.115$$

$$\lambda_{cd} = \lambda_{qd} - \frac{1-\lambda_{qd}}{N_c\tan\phi} = 1.115 - \frac{1-1.155}{20.72(\tan 25)} = 1.099$$

$$\lambda_{\gamma d} = 1$$

So,

$$q_u = (48)(20.72)(1.257)(1.099) + (0.6)(18)(10.66)(1.233)(1.115)$$

$$+ \frac{1}{2}(18)(0.6)(10.88)(0.8)(1)$$

$$= 1373.9 + 163.96 + 47 \approx \textbf{1585 kN/m}^2$$

Part b:
From Table 2.3 for $\phi = 25°$, $N_c = 20.72$, $N_q = 10.66$, and $N_\gamma = 6.77$. Now referring to Table 2.6, Meyerhof's shape and depth factors are as follows:

$$\lambda_{cs} = 1+0.2\left(\frac{B}{L}\right)\tan^2\left(45+\frac{\phi}{2}\right) = 1+(0.2)\left(\frac{0.6}{1.2}\right)\tan^2\left(45+\frac{25}{2}\right) = 1.246$$

$$\lambda_{qs} = \lambda_{\gamma s} = 1+0.1\left(\frac{B}{L}\right)\tan^2\left(45+\frac{\phi}{2}\right) = 1+0.1\left(\frac{0.6}{1.2}\right)\tan^2\left(45+\frac{25}{2}\right) = 1.123$$

$$\lambda_{cd} = 1+0.2\left(\frac{D_f}{B}\right)\tan\left(45+\frac{\phi}{2}\right) = 1+0.2\left(\frac{0.6}{0.6}\right)\tan\left(45+\frac{25}{2}\right) = 1.314$$

$$\lambda_{qd} = \lambda_{\gamma d} = 1+0.1\left(\frac{D_f}{B}\right)\tan\left(45+\frac{\phi}{2}\right) = 1+0.1\left(\frac{0.6}{0.6}\right)\tan\left(45+\frac{25}{2}\right) = 1.157$$

So,

$$q_u = (48)(20.72)(1.246)(1.314)+(0.6)(18)(10.66)(1.123)(1.157)$$

$$+\frac{1}{2}(18)(0.6)(6.77)(1.123)(1.157)$$

$$= 1628.37+149.6+47.7 \approx \mathbf{1826\ kN/m^2}$$

2.10   EFFECT OF SOIL COMPRESSIBILITY

In section 2.3 the ultimate bearing capacity equations proposed by Terzaghi[1] for local shear failure were given [equations (2.39)–(2.41)]. Also, suggestions by Vesic[4] shown in equations (2.42) and (2.43) address the problem of soil compressibility and its effect on soil bearing capacity. In order to account for soil compressibility Vesic[4] proposed the following modifications to equation (2.82), or

$$q_u = cN_c\lambda_{cs}\lambda_{cd}\lambda_{cc} + qN_q\lambda_{qs}\lambda_{qd}\lambda_{qc} + \tfrac{1}{2}\gamma BN_\gamma\lambda_{\gamma s}\lambda_{\gamma d}\lambda_{\gamma c} \qquad (2.83)$$

where
$\lambda_{cc}, \lambda_{qc}, \lambda_{\gamma c}$ = soil compressibility factors

The soil compressibility factors were derived by Vesic[4] from the analogy of expansion of cavities.[30] According to this theory, in order to calculate λ_{cc}, λ_{qc}, and $\lambda_{\gamma c}$, the following steps should be taken:

1. Calculate the rigidity index I_r of the soil (approximately at a depth of $B/2$ below the bottom of the foundation), or

$$I_r = \frac{G}{c+q\tan\phi} \qquad (2.84)$$

where
G = shear modulus of the soil

ϕ = soil friction angle

q = effective overburden pressure at the level of the foundation

2. The critical rigidity index of the soil $I_{r(cr)}$ can be expressed as

$$I_{r(cr)} = \frac{1}{2}\left\{ \exp\left[\left(3.3 - 0.45\frac{B}{L} \right) \cot\left(45 - \frac{\phi}{2} \right) \right] \right\} \tag{2.85}$$

3. If $I_r \geq I_{r(cr)}$, then use λ_{cc}, λ_{qc}, and $\lambda_{\gamma c}$ equal to one. However, if $I_r < I_{r(cr)}$,

$$\lambda_{\gamma c} = \lambda_{qc} = \exp\left\{ \left[-4.4 + 0.6\left(\frac{B}{L} \right) \right] \tan\phi + \frac{(3.07\sin\phi)(\log 2I_r)}{1 + \sin\phi} \right\} \tag{2.86}$$

For $\phi = 0$,

$$\lambda_{cc} = 0.32 + 0.12\frac{B}{L} + 0.6\log I_r \tag{2.87}$$

For other friction angles,

$$\lambda_{cc} = \lambda_{qc} - \frac{1 - \lambda_{qc}}{N_c \tan\phi} \tag{2.88}$$

Figures 2.22 and 2.23 show the variations of $\lambda_{\gamma c} = \lambda_{qc}$ [Eq. (2.86)] with ϕ and I_r.

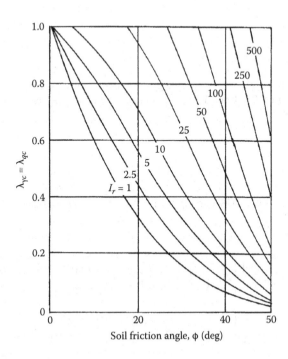

FIGURE 2.22 Variation of $\lambda_{\gamma c} = \lambda_{qc}$ with ϕ and I_r for square foundation ($B/L = 1$).

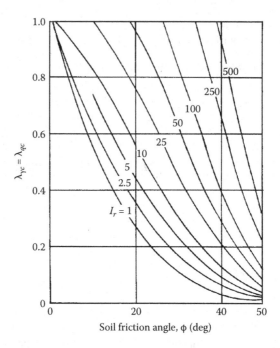

FIGURE 2.23 Variation of $\lambda_{\gamma c} = \lambda_{qc}$ with ϕ and I_r for foundation with $L/B > 5$.

EXAMPLE 2.2

Refer to Example 2.1a. For the soil, given: modulus of elasticity $E = 620$ kN/m²; Poisson's ratio $v = 0.3$. Considering the compressibility factors, determine the ultimate bearing capacity.

Solution

$$I_r = \frac{G}{c + q\tan\phi} = \frac{E}{2(1+v)(c+q\tan\phi)} = \frac{620}{2(1+0.3)[48+(18\times0.6)\tan25]} = 4.5$$

From equation (2.85):

$$I_{r(cr)} = \frac{1}{2}\left\{\exp\left[\left(3.3-0.45\frac{B}{L}\right)\cot\left(45-\frac{\phi}{2}\right)\right]\right\}$$

$$= \frac{1}{2}\left\{\exp\left[\left(3.3-0.45\times\frac{0.6}{1.2}\right)\cot\left(45-\frac{25}{2}\right)\right]\right\} = 62.46$$

Since $I_{r(cr)} > I_r$, use λ_{cc}, λ_{qc}, and $\lambda_{\gamma c}$ relationships from equations (2.86) and (2.88):

$$\lambda_{\gamma c} = \lambda_{qc} = \exp\left\{\left[-4.4 + 0.6\left(\frac{B}{L}\right)\right]\tan\phi + \frac{(3.07\sin\phi)(\log 2I_r)}{1 + \sin\phi}\right\}$$

$$= \exp\left\{\left[-4.4 + 0.6\left(\frac{0.6}{1.2}\right)\right]\tan 25 + \frac{(3.07\sin 25)\log(2 \times 4.5)}{1 + \sin 25}\right\} = 0.353$$

Also,

$$\lambda_c = \lambda_{qc} - \frac{1 - \lambda_{qc}}{N_c \tan\phi} = 0.353 - \frac{1 - 0.353}{20.72\tan 25} = 0.286$$

equation (2.83):

$$q_u = (48)(20.72)(1.257)(1.099)(0.286) + (0.6)(18)(10.66)(1.233)(1.115)(0.353)$$

$$+ \frac{1}{2}(18)(0.6)(10.88)(0.8)(1)(0.353)$$

$$= 392.94 + 55.81 + 16.59 \approx \mathbf{465.4\ kN/m^2}$$

2.11 BEARING CAPACITY OF FOUNDATIONS ON ANISOTROPIC SOILS

2.11.1 FOUNDATION ON SAND (c = 0)

Most natural deposits of cohesionless soil have an inherent anisotropic structure due to their nature of deposition in horizontal layers. The initial deposition of the granular soil and the subsequent compaction in the vertical direction cause the soil particles to take a preferred orientation. For a granular soil of this type Meyerhof suggested that, if the *direction of application of deviator stress* makes an angle i with the direction of deposition of soil (Figure 2.24), then the soil friction angle ϕ can be

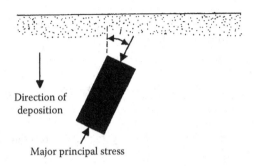

Direction of deposition

Major principal stress

FIGURE 2.24 Anisotropy in sand deposit.

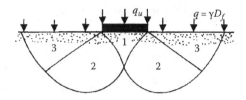

FIGURE 2.25 Continuous rough foundation on anisotropic sand deposit.

approximated in a form

$$\phi = \phi_1 - (\phi_1 - \phi_2)\left(\frac{i°}{90°}\right) \tag{2.89}$$

where
ϕ_1 = soil friction angle with $i = 0°$
ϕ_2 = soil friction angle with $i = 90°$

Figure 2.25 shows a continuous (strip) rough foundation on an anisotropic sand deposit. The failure zone in the soil at ultimate load is also shown in the figure. In the triangular zone (zone 1) the soil friction angle will be $\phi = \phi_1$; however, the magnitude of ϕ will vary between the limits of ϕ_1 and ϕ_2 in zone 2. In zone 3 the effective friction angle of the soil will be equal to ϕ_2. Meyerhof[31] suggested that the ultimate bearing capacity of a continuous foundation on anisotropic sand could be calculated by assuming an equivalent friction angle $\phi = \phi_{eq}$, or

$$\phi_{eq} = \frac{(2\phi_1 + \phi_2)}{3} = \frac{(2 + m)\phi_1}{3} \tag{2.90}$$

where

$$m = \text{friction ratio} = \frac{\phi_2}{\phi_1} \tag{2.91}$$

Once the equivalent friction angle is determined, the ultimate bearing capacity for vertical loading conditions on the foundation can be expressed as (neglecting the depth factors)

$$q_u = qN_{q(eq)}\lambda_{qs} + \tfrac{1}{2}\gamma BN_{\gamma(eq)}\lambda_{\gamma s} \tag{2.92}$$

where
$N_{q(eq)}$, $N_{\gamma(eq)}$ = equivalent bearing capacity factors corresponding to the friction angle $\phi = \phi_{eq}$

In most cases the value of ϕ_1 will be known. Figures 2.26 and 2.27 present the plots of $N_{\gamma(eq)}$ and $N_{q(eq)}$ in terms of m and ϕ_1. Note that the soil friction angle $\phi = \phi_{eq}$

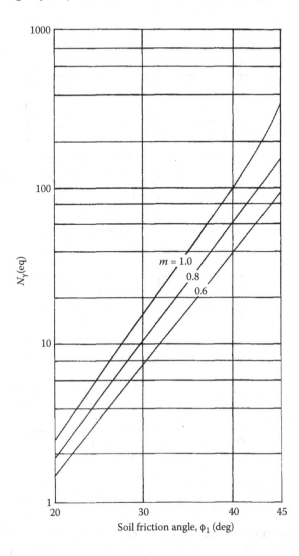

FIGURE 2.26 Variation of $N_{\gamma(eq)}$ [equation (2.92)].

was used in equations (2.66) and (2.72) to prepare the graphs. So, combining the relationships for shape factors (Table 2.5) given by DeBeer,[19]

$$q_u = qN_{q(eq)}\left[1+\left(\frac{B}{L}\right)\tan\phi_{eq}\right]+\frac{1}{2}\gamma BN_{\gamma(eq)}\left[1-0.4\left(\frac{B}{L}\right)\right] \qquad (2.93)$$

2.11.2 FOUNDATIONS ON SATURATED CLAY ($\phi = 0$ CONCEPT)

As in the case of sand discussed above, saturated clay deposits also exhibit aniso-tropic undrained shear strength properties. Figures 2.28a and b show the nature of

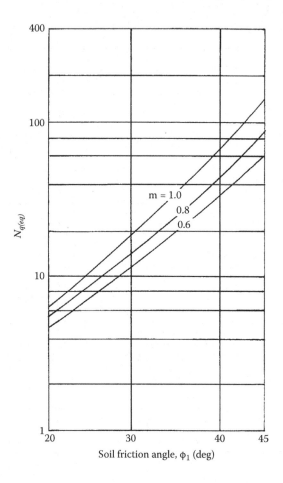

FIGURE 2.27 Variation of $N_{q(eq)}$ [equation (2.92)].

variation of the undrained shear strength of clays c_u with respect to the direction of principal stress application.[32] Note that the undrained shear strength plot shown in Figure 2.28b is elliptical; however, the center of the ellipse does not match the origin. The geometry of the ellipse leads to the equation

$$\frac{b}{a} = \frac{c_{u(i=45°)}}{\sqrt{(c_{uV})(c_{uH})}} \qquad (2.94)$$

where

c_{uV} = undrained shear strength with $i = 0°$
c_{uH} = undrained shear strength with $i = 90°$

A continuous foundation on a saturated clay layer ($\phi = 0$) whose directional strength variation follows equation (2.94) is shown in Figure 2.28c. The failure

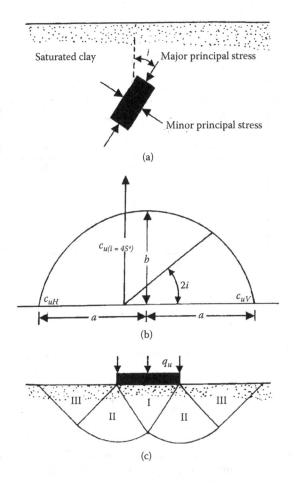

FIGURE 2.28 Bearing capacity of continuous foundation on anisotropic saturated clay.

surface in the soil at ultimate load is also shown in the figure. Note that, in zone I, the major principal stress direction is vertical. The direction of the major principal stress is horizontal in zone III; however, it gradually changes from vertical to horizontal in zone II. Using the stress characteristic solution, Davis and Christian[32] determined the bearing capacity factor $N_{c(i)}$ for the foundation. For a surface foundation,

$$q_u = N_{c(i)} \left(\frac{c_{uV} + c_{uH}}{2} \right) \tag{2.95}$$

The variation of $N_{c(i)}$ with the ratio of a/b (Figure 2.28b) is shown in Figure 2.29. Note that, when $a = b$, $N_{c(i)}$ becomes equal to $N_c = 5.14$ [isotropic case; equation (2.67)].

In many practical conditions, the magnitudes of c_{uV} and c_{uH} may be known but not the magnitude of $c_{u(i\,=\,45°)}$. If such is the case, the magnitude of a/b [equation

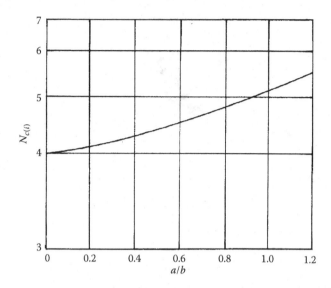

FIGURE 2.29 Variation of $N_{c(i)}$ with a/b based on the analysis of Davis and Christian.

(2.94)] cannot be determined. For such conditions, the following approximation may be used:

$$q_u \approx 0.9 \underbrace{N_c}_{=5.14} \left(\frac{c_{uV} + c_{uH}}{2} \right) \qquad (2.96)$$

The preceding equation was suggested by Davis and Christian,[32] and it is based on the undrained shear strength results of several clays. So, in general, for a rectangular foundation with vertical loading condition,

$$q_u = N_{c(i)} \left[\frac{c_{uV} + c_{uH}}{2} \right] \lambda_{cs} \lambda_{cd} + q N_q \lambda_{qs} \lambda_{qd} \qquad (2.97)$$

For $\phi = 0$ condition, $N_q = 1$ and $q = \gamma D_f$. So,

$$q_u = N_{c(i)} \left(\frac{c_{uV} + c_{uH}}{2} \right) \lambda_{cs} \lambda_{cd} + q D_f \lambda_{qs} \lambda_{qd} \qquad (2.98)$$

The desired relationships for the shape and depth factors can be taken from Table 2.6 and the magnitude of q_u can be estimated.

2.11.3 FOUNDATIONS ON $c-\phi$ SOIL

The ultimate bearing capacity of a continuous shallow foundation supported by anisotropic $c-\phi$ soil was studied by Reddy and Srinivasan[33] using the method of characteristics. According to this analysis the shear strength of a soil can be given as

$$s = \sigma' \tan\phi + c$$

It is assumed, however, that the soil is anisotropic only with respect to cohesion. As mentioned previously in this section, the direction of the major principal stress (with respect to the vertical) along a slip surface located below the foundation changes. In anisotropic soils, this will induce a change in the shearing resistance to the bearing capacity failure of the foundation. Reddy and Srinivasan[33] assumed the directional variation of c at a given depth z below the foundation as (Figure 2.30a)

$$c_{i(z)} = c_{H(z)} + [c_{V(z)} - c_{H(z)}]\cos^2 i \tag{2.99}$$

where

$c_{i(z)}$ = cohesion at a depth z when the major principal stress is inclined at an angle i to the vertical (Figure 2.30b)

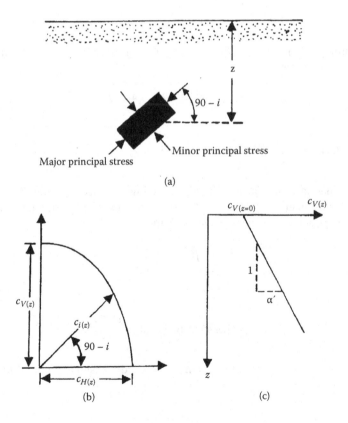

FIGURE 2.30 Anisotropic clay soil—assumptions for bearing capacity evaluation.

$c_{V(z)}$ = cohesion at depth z for $i = 0°$
$c_{H(z)}$ = cohesion at depth z for $i = 90°$

The preceding equation is of the form suggested by Casagrande and Carrillo.[34]

Figure 2.30b shows the nature of variation of $c_{i(z)}$ with i. The anisotropy coefficient K is defined as the ratio of $c_{V(z)}$ to $c_{H(z)}$:

$$K = \frac{c_{V(z)}}{c_{H(z)}} \qquad (2.100)$$

In *overconsolidated* soils, K is less than one; for *normally consolidated* soils, the magnitude of K is greater than one.

For many consolidated soils, the cohesion increases linearly with depth (Figure 2.30c). Thus,

$$c_{V(z)} = c_{V(z=0)} + \alpha' z \qquad (2.101)$$

where

$c_{V(z)}$, $c_{V(z=0)}$ = cohesion in the vertical direction (that is, $i = 0$) at depths of z and
$\qquad z = 0$, respectively
α' = the rate of variation with depth z

According to this analysis, the ultimate bearing capacity of a continuous foundation may be given as

$$q_u = c_{V(z=0)} N_{c(i')} + q N_{q(i')} + \tfrac{1}{2} \gamma B N_{\gamma(i')} \qquad (2.102)$$

where

$N_{c(i')}, N_{q(i')}, N_{\gamma(i')}$ = bearing capacity factors
$q = \gamma D_f$

This equation is similar to Terzaghi's bearing capacity equation for continuous foundations [equation (2.31)]. The bearing capacity factors are functions of the parameters β_c and K. The term β_c can be defined as

$$\beta_c = \frac{\alpha' l}{c_{V(z=0)}} \qquad (2.103)$$

where

$$l = \text{characteristic length} = \frac{c_{V(z=0)}}{\gamma} \qquad (2.104)$$

Furthermore, $N_{c(i')}$ is also a function of the nondimensional width of the foundation, B':

$$B' = \frac{B}{l} \qquad (2.105)$$

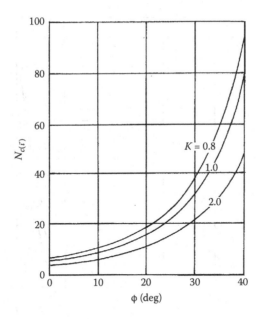

FIGURE 2.31 Reddy and Srinivasan's bearing capacity factor, $N_{c(i')}$—influence of K ($\beta_c = 0$).

The variations of the bearing capacity factors with β_c, B', ϕ, and K determined using the method of analysis by Reddy and Srinivasan[33] are shown in Figures 2.31 to 2.36. This study shows that the rupture surface in soil at ultimate load extends to a smaller distance below the bottom of the foundation for the case where the anisotropic coefficient K is greater than one. Also, when K changes from one to two with $\alpha' = 0$, the magnitude of $N_{c(i')}$ is reduced by about 30%–40%.

EXAMPLE 2.3

Estimate the ultimate bearing capacity q_u of a continuous foundation with the following: $B = 3$ m; $c_{V(z=0)} = 12$ kN/m²; $\alpha' = 3.9$ kN/m²/m; $D_f = 1$ m; $\gamma = 17.29$ kN/m³; $\phi = 20°$. Assume $K = 2$.

Solution

From equation (2.104):

$$\text{Characteristic length, } l = \frac{c_{V(z=0)}}{\gamma} = \frac{12}{17.29} = 0.69$$

$$\text{Nondimensional width, } B' = \frac{B}{l} = \frac{3}{0.69} = 4.34$$

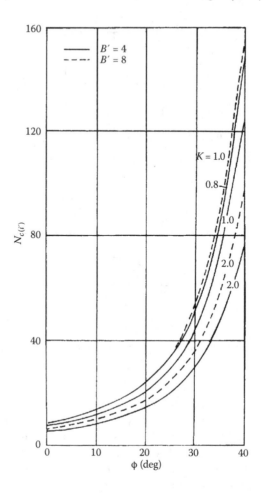

FIGURE 2.32 Reddy and Srinivasan's bearing capacity factor, $N_{c(i')}$—influence of K ($\beta_c = 0.2$).

Also,

$$\beta_c = \frac{\alpha'l}{c_{V(z=0)}} = \frac{(4.34)(0.69)}{12} = 0.25$$

Now, referring to Figures 2.32, 2.33, 2.35, and 2.36 for $\phi = 20°$, $\beta_c = 0.25$, $K = 2$, and $B' = 4.34$ (by interpolation),

$$N_{c(i')} \approx 14.5; \; N_{q(i')} \approx 6, \; \text{and} \; N_{\gamma(i')} \approx 4$$

From equation (2.102),

$$q_u = c_{V(z=0)}N_{c(i')} + qN_{q(i')} + \tfrac{1}{2}\gamma BN_{\gamma(i')} = (12)(14.5) + (1)(17.29)(6)$$
$$+ \tfrac{1}{2}(17.29)(3)(4) \approx \textbf{381 kN/m}^2$$

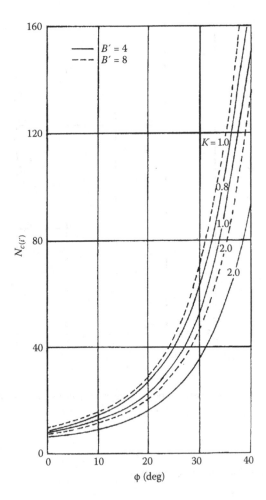

FIGURE 2.33 Reddy and Srinivasan's bearing capacity factor, $N_{c(i')}$—influence of K ($\beta_c = 0.4$).

2.12 ALLOWABLE BEARING CAPACITY WITH RESPECT TO FAILURE

Allowable bearing capacity for a given foundation may be (a) to protect the founda-tion against a bearing capacity failure, or (b) to ensure that the foundation does not undergo undesirable settlement. There are three definitions for the allowable capacity with respect to a bearing capacity failure.

2.12.1 GROSS ALLOWABLE BEARING CAPACITY

The gross allowable bearing capacity is defined as

$$q_{all} = \frac{q_u}{FS} \tag{2.106}$$

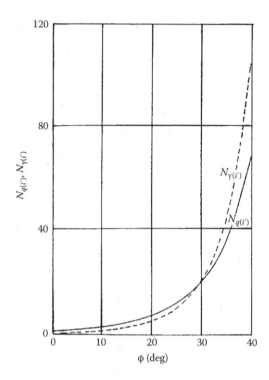

FIGURE 2.34 Reddy and Srinivasan's bearing capacity factors, $N_{\gamma(i')}$ and $N_{q(i')}$—influence of K ($\beta_c = 0$).

where

q_{all} = gross allowable bearing capacity
FS = factor of safety

In most cases a factor of safety of 3 to 4 is generally acceptable.

2.12.2 NET ALLOWABLE BEARING CAPACITY

The net ultimate bearing capacity is defined as the ultimate load per unit area of the foundation that can be supported by the soil in excess of the pressure caused by the surrounding soil at the foundation level. If the difference between the unit weight of concrete used in the foundation and the unit weight of the surrounding soil is assumed to be negligible, then

$$q_{u(net)} = q_u - q \tag{2.107}$$

where

$q = \gamma D_f$
$q_{u(net)}$ = net ultimate bearing capacity

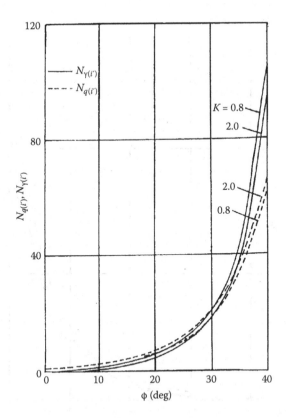

FIGURE 2.35 Reddy and Srinivasan's bearing capacity factors, $N_{\gamma(i')}$ and $N_{q(i')}$—influence of K ($\beta_c = 0.2$).

The net allowable bearing capacity can now be defined as

$$q_{\text{all(net)}} = \frac{q_{u(\text{net})}}{FS} \tag{2.108}$$

A factor of safety of 3 to 4 in the preceding equation is generally considered satisfactory.

2.12.3   ALLOWABLE BEARING CAPACITY WITH RESPECT TO SHEAR FAILURE [$q_{\text{all(shear)}}$]

For this case a factor of safety with respect to shear failure $FS_{(\text{shear})}$, which may be in the range of 1.3–1.6, is adopted. In order to evaluate $q_{\text{all(shear)}}$, the following procedure may be used:

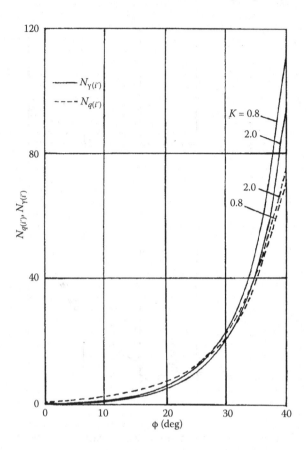

FIGURE 2.36 Reddy and Srinivasan's bearing capacity factors, $N_{\gamma(i')}$ and $N_{q(i')}$—influence of K ($\beta_c = 0.4$).

1. Determine the developed cohesion c_d and the developed angle of friction ϕ_d as

$$c_d = \frac{c}{FS_{(shear)}} \tag{2.109}$$

$$\phi_d = \tan^{-1}\left[\frac{\tan\phi}{FS_{(shear)}}\right] \tag{2.110}$$

2. The gross and net ultimate allowable bearing capacities with respect to shear failure can now be determined as [equation (2.82)]

$$q_{all(shear)-gross} = c_d N_c \lambda_{cs} \lambda_{cd} + q N_q \lambda_{qs} \lambda_{qd} + \tfrac{1}{2}\gamma B N_\gamma \lambda_{\gamma s} \lambda_{\gamma d} \tag{2.111}$$

$$q_{\text{all(shear)}-\text{net}} = q_{\text{all(shear)}-\text{gross}} - q = c_d N_c \lambda_{cs} \lambda_{cd} + q(N_q - 1)\lambda_{qs}\lambda_{qd} + \tfrac{1}{2}\gamma B N_\gamma \lambda_{\gamma s}\lambda_{\gamma d}$$

$$(2.112)$$

where
N_c, N_q, and N_γ = bearing capacity factors for friction angle ϕ_d

EXAMPLE 2.4

Refer to Example 2.1, problem a.

    a. Determine the gross allowable bearing capacity. Assume $FS = 4$.
    b. Determine the net allowable bearing capacity. Assume $FS = 4$.
    c. Determine the gross and net allowable bearing capacities with respect to shear failure. Assume $FS_{(\text{shear})} = 1.5$.

Solution

Part a
From Example 2.1, problem a, $q_u = 1585$ kN/m²

$$q_{\text{all}} = \frac{q_u}{FS} = \frac{1585}{4} \approx \textbf{396.25 kN/m}^2$$

Part b

$$q_{\text{all(net)}} = \frac{q_u - q}{FS} = \frac{1585 - (0.6)(18)}{4} \approx \textbf{393.55 kN/m}^2$$

Part c

$$c_d = \frac{c}{FS_{(\text{shear})}} = \frac{48}{1.5} = 32 \text{ kN/m}^2$$

$$\phi_d = \tan^{-1}\left[\frac{\tan\phi}{FS_{(\text{shear})}}\right]\tan^{-1}\left[\frac{\tan 25}{1.5}\right] = 17.3°$$

For $\phi_d = 17.3°$, $N_c = 12.5$, $N_q = 4.8$ (Table 2.3), and $N_\gamma = 3.6$ (Table 2.4),

$$\lambda_{cs} = 1 + \left(\frac{N_q}{N_c}\right)\left(\frac{B}{L}\right) = 1 + \left(\frac{4.8}{12.5}\right)\left(\frac{0.6}{1.2}\right) = 1.192$$

$$\lambda_{qs} = 1 + \left(\frac{B}{L}\right)\tan\phi_d = 1 + \left(\frac{0.6}{1.2}\right)\tan 17.3 = 1.156$$

$$\lambda_{\gamma s} = 1 - 0.4\left(\frac{B}{L}\right) = 1 - 0.4\left(\frac{0.6}{1.2}\right) = 0.8$$

$$\lambda_{cd} = \lambda_{qd} - \frac{1-\lambda_{qd}}{N_c \tan \phi_d} = 1.308 - \frac{1-1.308}{12.5 \tan 17.3} = 1.387$$

$$\lambda_{qd} = 1 + 2 \tan \phi_d (1-\sin \phi_d)^2 \left(\frac{D_f}{B} \right) = 1 + (2)(\tan 17.3)(1-\sin 17.3)^2 \left(\frac{0.6}{0.6} \right) = 1.308$$

$$\lambda_{\gamma d} = 1$$

From equation (2.111)

$$q_{\text{all(shear)-gross}} = c_d N_c \lambda_{cs} \lambda_{cd} + q N_q \lambda_{qs} \lambda_{qd} + \tfrac{1}{2} \gamma B N_\gamma \lambda_{\gamma s} \lambda_{\gamma d}$$

$$= (32)(12.5)(1.192)(1.387) + (0.6)(18)(4.8)(1.156)(1.308)$$

$$+ \tfrac{1}{2}(18)(0.6)(3.6)(0.8)(1)$$

$$= 661.3 + 78.4 + 15.6 = 755.3 \text{ kN/m}^2$$

From equation (2.112):

$$q_{\text{all(shear)-net}} = 761.5 - q = 755.3 - (0.6)(18) \approx \mathbf{744.5 \text{ kN/m}^2}$$

2.13 INTERFERENCE OF CONTINUOUS FOUNDATIONS IN GRANULAR SOIL

In earlier sections of this chapter, theories relating to the ultimate bearing capacity of single rough continuous foundations supported by a homogeneous soil medium extending to a great depth were discussed. However, if foundations are placed close to each other with similar soil conditions, the ultimate bearing capacity of each foundation may change due to the interference effect of the failure surface in the soil. This was theoretically investigated by Stuart[35] for *granular soils*. The results of this study are summarized in this section. Stuart[35] assumed the geometry of the rupture surface in the soil mass to be the same as that assumed by Terzaghi (Figure 2.1). According to Stuart, the following conditions may arise (Figure 2.37):

CASE 1 (FIGURE 2.37a)

If the center-to-center spacing of the two foundations is $x \geq x_1$, the rupture surface in the soil under each foundation will not overlap. So the ultimate bearing capacity of each continuous foundation can be given by Terzaghi's equation [equation (2.31)]. For $c = 0$,

$$q_u = q N_q + \tfrac{1}{2} \gamma B N_\gamma \tag{2.113}$$

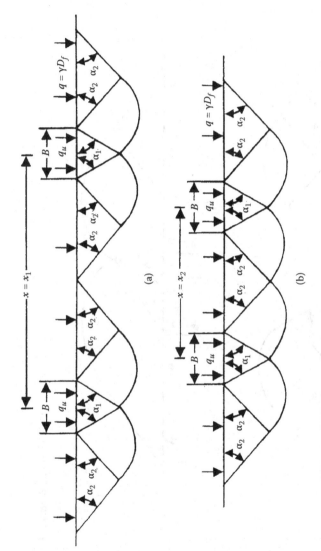

FIGURE 2.37 Assumptions for the failure surface in granular soil under two closely spaced rough continuous foundations. *Note:* $\alpha_1 = \phi$, $\alpha_2 = 45 - \phi/2$, $\alpha_3 = 180 - \phi$.

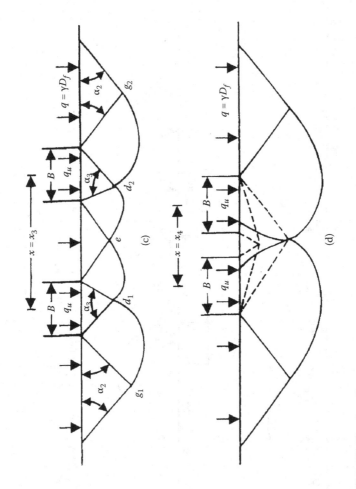

FIGURE 2.37 (Continued).

where

N_q, N_γ = Terzaghi's bearing capacity factors (Table 2.1)

CASE 2 (FIGURE 2.37b)

If the center-to-center spacing of the two foundations ($x = x_2 < x_1$) are such that the Rankine passive zones just overlap, then the magnitude of q_u will still be given by equation (2.113). However, the foundation settlement at ultimate load will change (compared to the case of an isolated foundation).

CASE 3 (FIGURE 2.37c)

This is the case where the center-to-center spacing of the two continuous foundations is $x = x_3 < x_2$. Note that the triangular wedges in the soil under the foundation make angles of $180° - 2\phi$ at points d_1 and d_2. The arcs of the logarithmic spirals $d_1 g_1$ and $d_1 e$ are tangent to each other at point d_1. Similarly, the arcs of the logarithmic spirals $d_2 g_2$ and $d_2 e$ are tangent to each other at point d_2. For this case, the ultimate bearing capacity of each foundation can be given as ($c = 0$)

$$q_u = q N_q \zeta_q + \tfrac{1}{2} \gamma B N_\gamma \zeta_\gamma \tag{2.114}$$

where

ξ_q, ξ_γ = efficiency ratios

The efficiency ratios are functions of x/B and soil friction angle ϕ. The theoretical variations of ξ_q and ξ_γ are given in Figures 2.38 and 2.39.

CASE 4 (FIGURE 2.37d)

If the spacing of the foundation is further reduced such that $x = x_4 < x_3$, blocking will occur and the pair of foundations will act as a single foundation. The soil between the individual units will form an inverted arch that travels down with the foundation as

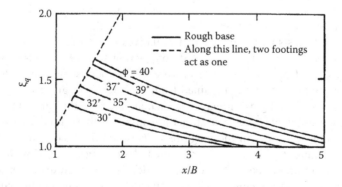

FIGURE 2.38 Stuart's interference factor ξ_q.

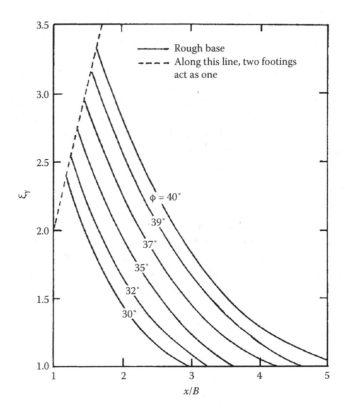

FIGURE 2.39 Stuart's interference factor ξ_γ.

the load is applied. When the two foundations touch, the zone of arching disappears and the system behaves as a single foundation with a width equal to $2B$. The ultimate bearing capacity for this case can be given by equation (2.113), with B being replaced by $2B$ in the third term.

Das and Larbi-Cherif[36] conducted laboratory model tests to determine the interference efficiency ratios ξ_q and ξ_γ of two rough continuous foundations resting on sand extending to a great depth. The sand used in the model tests was highly angular, and the tests were conducted at a relative density of about 60%. The angle of friction ϕ at this relative density of compaction was 39°. Load-displacement curves obtained from the model tests were of the local shear type. The experimental variations of ξ_q and ξ_γ obtained from these tests are given in Figures 2.40 and 2.41. From these figures it may be seen that, although the general trend of the experimental efficiency ratio variations is similar to those predicted by theory, there is a large variation in the magnitudes between the theory and experimental results. Figure 2.42 shows the experimental variations of S_u/B with x/B ($S_u =$ settlement at ultimate load). The elastic settlement of the foundation decreases with the increase in the center-to-center spacing of the foundation and remains constant at $x >$ about $4B$.

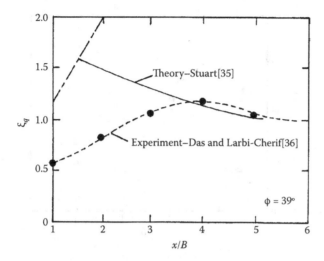

FIGURE 2.40 Comparison of experimental and theoretical ξ_q.

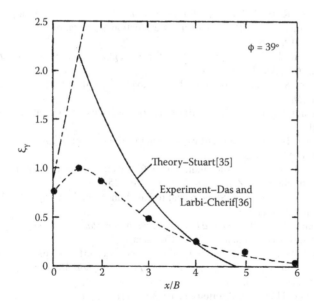

FIGURE 2.41 Comparison of experimental and theoretical ξ_γ.

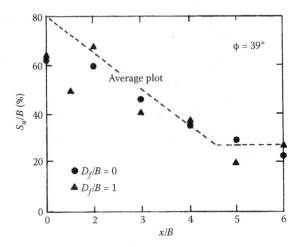

FIGURE 2.42 Variation of experimental elastic settlement (S_u/B) with center-to-center spacing of two continuous rough foundations.

REFERENCES

1. Terzaghi, K. 1943. *Theoretical soil mechanics*. New York: John Wiley.
2. Kumbhojkar, A. S. 1993. Numerical evaluation of Terzaghi's N_γ. *J. Geotech. Eng.*, ASCE, 119(3): 598.
3. Krizek, R. J. 1965. Approximation for Terzaghi's bearing capacity. *J. Soil Mech. Found. Div.*, ASCE, 91(2): 146.
4. Vesic, A. S. 1973. Analysis of ultimate loads of shallow foundations. *J. Soil Mech. Found. Div.*, ASCE, 99(1): 45.
5. Meyerhof, G. G. 1951. The ultimate bearing capacity of foundations. *Geotechnique.* 2: 301.
6. Reissner, H. 1924. Zum erddruckproblem, in *Proc., First Intl. Conf. Appl. Mech.*, Delft, The Netherlands, 295.
7. Prandtl, L. 1921. Uber die eindringungs-festigkeit plastisher baustoffe und die festigkeit von schneiden. *Z. Ang. Math. Mech.* 1(1): 15.
8. Meyerhof, G. G. 1963. Some recent research on the bearing capacity of foundations. *Canadian Geotech. J.* 1(1): 16.
9. Hansen, J. B. 1970. *A revised and extended formula for bearing capacity.* Bulletin No. 28, Danish Geotechnical Institute, Copenhagen.
10. Caquot, A., and J. Kerisel. 1953. Sue le terme de surface dans le calcul des fondations en milieu pulverulent, in *Proc., III Intl. Conf. Soil Mech. Found. Eng.*, Zurich, Switzerland, 1: 336.
11. Lundgren, H., and K. Mortensen. 1953. Determination by the theory of plasticity of the bearing capacity of continuous footings on sand, in *Proc., III Intl. Conf. Mech. Found. Eng.*, Zurich, Switzerland, 1: 409.
12. Chen, W. F. 1975. *Limit analysis and soil plasticity.* New York: Elsevier Publishing Co.
13. Drucker, D. C., and W. Prager. 1952. Soil mechanics and plastic analysis of limit design. *Q. Appl. Math.* 10: 157.

14. Biarez, J., M. Burel, and B. Wack. 1961. Contribution à l'étude de la force portante des fondations, in *Proc., V Intl. Conf. Soil Mech. Found. Eng.,* Paris, France, 1: 603.
15. Booker, J. R. 1969. Application of theories of plasticity to cohesive frictional soils. Ph.D. thesis, Sydney University, Australia.
16. Poulos, H. G., J. P. Carter, and J. C. Small. 2001. Foundations and retaining structures—research and practice, in *Proc. 15th Intl. Conf. Soil Mech. Found. Eng.,* Istanbul, Turkey, 4, A. A. Balkema, Rotterdam, 2527.
17. Kumar, J. 2003. N_γ for rough strip footing using the method of characteristics. *Canadian Geotech. J.* 40(3): 669.
18. Michalowski, R. L. 1997. An estimate of the influence of soil weight on bearing capacity using limit analysis. *Soils and Foundations.* 37(4): 57.
19. Hjiaj, M., A. V. Lyamin, and S. W. Sloan. 2005. Numerical limit analysis solutions for the bearing capacity factor N_γ. *Int. J. of Soils and Struc.* 43: 1681.
20. Martin, C. M. 2005. Exact bearing capacity calculations using the method of characteristics. *Proc., 11th Int. Conf. IACMAG,* Turin, 4: 441.
21. Salgado, R. 2008. *The engineering of foundations.* New York: McGraw-Hill.
22. Ko, H. Y., and L. W. Davidson. 1973. Bearing capacity of footings in plane strain. *J. Soil Mech. Found. Div.,* ASCE, 99(1): 1.
23. Hu, G. G. Y. 1964. Variable-factors theory of bearing capacity. *J. Soil Mech. Found. Div.,* ASCE, 90(4): 85.
24. Balla, A. 1962. Bearing capacity of foundations. *J. Soil Mech. Found. Div.,* ASCE, 88(5): 13.
25. DeBeer, E. E. 1965. Bearing capacity and settlement of shallow foundations on sand, in *Bearing capacity and settlement of foundations,* Proceedings of a symposium held at Duke University: 15.
26. Muhs, E. 1963. Ueber die zulässige Belastung nicht bindigen Böden—Mitteilungen der Degebo—Berlin, Heft: 16.
27. DeBeer, E. E. 1970. Experimental determination of the shape factors of sand. *Geotechnique.* 20(4): 307.
28. Michalowski, R. L. 1997. An estimate of the influence of soil weight on bearing capacity using limit analysis. *Soils and Foundations.* 37(4).
29. Salgado, R., A. V. Lyamin, S. W. Sloan, and H. S. Yu. 2004. Two- and three-dimensional bearing capacity of foundations in clay. *Geotechnique.* 54(5).
30. Vesić, A. 1963. *Theoretical studies of cratering mechanisms affecting the stability of cratered slopes.* Final Report, Project No. A-655, Engineering Experiment Station, Georgia Institute of Technology, Atlanta, GA.
31. Meyerhof, G. G. 1978. Bearing capacity of anisotropic cohesionless soils. *Canadian Geotech. J.* 15(4): 593.
32. Davis, E., and J. T. Christian. 1971. Bearing capacity of anisotropic cohesive soil. *J. Soil Mech. Found. Div.,* ASCE, 97(5): 753.
33. Reddy, A. S., and R. J. Srinivasan. 1970. Bearing capacity of footings on anisotropic soils. *J. Soil Mech. Found. Div.,* ASCE, 96(6): 1967.
34. Casagrande, A., and N. Carrillo. 1944. Shear failure in anisotropic materials, in *Contribution to soil mechanics 1941–53,* Boston Society of Civil Engineers: 122.
35. Stuart, J. G. 1962. Interference between foundations with special reference to surface footing on sand. *Geotechnique.* 12(1): 15.
36. Das, B. M., and S. Larbi-Cherif. 1983. Bearing capacity of two closely spaced shallow foundations on sand. *Soils and Foundations.* 23(1): 1.

3 Ultimate Bearing Capacity under Inclined and Eccentric Loads

3.1 INTRODUCTION

Due to bending moments and horizontal thrusts transferred from the superstructure, shallow foundations are often subjected to eccentric and inclined loads. Under such circumstances the ultimate bearing capacity theories presented in Chapter 2 need some modification, and this is the subject of discussion in this chapter. The chapter is divided into two major parts. The first part discusses the ultimate bearing capacities of shallow foundations subjected to centric inclined loads, and the second part is devoted to the ultimate bearing capacity under eccentric loading.

3.2 FOUNDATIONS SUBJECTED TO INCLINED LOAD

3.2.1 Meyerhof's Theory (Continuous Foundation)

In 1953, Meyerhof[1] extended his theory for ultimate bearing capacity under vertical loading (section 2.4) to the case with inclined load. Figure 3.1 shows the plastic zones in the soil near a rough continuous (strip) foundation with a small inclined load. The shear strength of the soil s is given as

$$s = c + \sigma' \tan \phi \qquad (3.1)$$

where

c = cohesion
σ' = effective vertical stress
ϕ = angle of friction

The inclined load makes an angle α with the vertical. It needs to be pointed out that Figure 3.1 is an extension of Figure 2.7. In Figure 3.1, abc is an elastic zone, bcd is a radial shear zone, and bde is a mixed shear zone. The normal and shear stresses on plane be are p_o and s_o, respectively. Also, the unit base adhesion is c_a'. The solution for the ultimate bearing capacity q_u can be expressed as

$$q_{u(v)} = q_u \cos \alpha = cN_c + p_o N_q + \tfrac{1}{2}\gamma B N_\gamma \qquad (3.2)$$

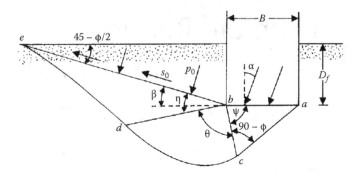

FIGURE 3.1 Plastic zones in soil near a foundation with an inclined load.

where
N_c, N_q, N_γ = bearing capacity factors for inclined loading condition
γ = unit weight of soil

Similar to equations (2.71), (2.59), and (2.70), we can write

$$q_{u(v)} = q_u \cos\alpha = q'_{u(v)} + q''_{u(v)} \tag{3.3}$$

where

$$q'_{u(v)} = cN_c + p_o N_q \quad \text{(for } \phi \neq 0, \ \gamma = 0, \ p_o \neq 0, \ c \neq 0) \tag{3.4}$$

and

$$q''_{u(v)} = \tfrac{1}{2}\gamma B N_\gamma \quad \text{(for } \phi \neq 0, \ \gamma \neq 0, \ p_o = 0, \ c = 0) \tag{3.5}$$

It was shown by Meyerhof[1] in equation (3.4) that

$$N_c = \left\{ \cot\phi \left[\frac{1+\sin\phi\sin(2\psi-\phi)}{1-\sin\phi\sin(2\eta+\phi)} e^{2\theta\tan\phi} - 1 \right] \right\} \tag{3.6}$$

$$N_c = \frac{1+\sin\phi\sin(2\psi-\phi)}{1-\sin\phi\sin(2\eta+\phi)} e^{2\theta\tan\phi} \tag{3.7}$$

Note that the horizontal component of the inclined load per unit area on the foundation q'_h cannot exceed the shearing resistance at the base, or

$$q'_{u(h)} \leq c_a + q'_{u(v)} \tan\delta \tag{3.8}$$

where
c_a = unit base adhesion
δ = unit base friction angle

In order to determine the minimum passive force per unit length of the foundation $P_{p\gamma(min)}$ (see Figure 2.11 for comparison) to obtain N_γ, one can take a numerical step-by-step approach as shown by Caquot and Kerisel[2] or a semi-graphical approach based on the logarithmic spiral method as shown by Meyerhof.[3] Note that the passive force $P_{p\gamma}$ acts at an angle ϕ with the normal drawn to the face bc of the elastic wedge abc (Figure 3.1). The relationship for N_γ is

$$N_\gamma = \frac{2P_{p\gamma(min)}}{\gamma B^2}\left[\frac{\sin^2\psi}{\cos(\psi-\phi)}+\cos(\psi-\phi)\right]-\frac{\sin\psi\cos(\psi-\phi)}{\cos\phi} \quad (\text{for } \alpha \leq \delta) \quad (3.9)$$

The ultimate bearing capacity expression given by equation (3.2) can also be expressed as

$$q_{u(v)} = q_u\cos\alpha = cN_{eq}+\tfrac{1}{2}\gamma BN_{\gamma q} \quad (3.10)$$

where

$N_{cq}, N_{\gamma q}$ = bearing capacity factors that are functions of the soil friction angle ϕ and the depth of the foundation D_f

For a purely cohesive soil ($\phi = 0$),

$$q_{u(v)} = q_u\cos\alpha = cN_{cq} \quad (3.11)$$

Figure 3.2 shows the variation of N_{cq} for a purely cohesive soil ($\phi = 0$) for various load inclinations α.

For cohesionless soils $c = 0$; hence, equation (3.10) gives

$$q_{u(v)} = q_u\cos\alpha = \tfrac{1}{2}\gamma BN_{\gamma q} \quad (3.12)$$

Figure 3.3 shows the variation of $N_{\gamma q}$ with α.

3.2.2 GENERAL BEARING CAPACITY EQUATION

The general ultimate bearing capacity equation for a rectangular foundation given by equation (2.82) can be extended to account for an inclined load and can be expressed as

$$q_u = cN_c\lambda_{cs}\lambda_{cd}\lambda_{ci}+qN_q\lambda_{qs}\lambda_{qd}\lambda_{qi}+\tfrac{1}{2}\gamma BN_\gamma\lambda_{\gamma s}\lambda_{\gamma d}\lambda_{\gamma i} \quad (3.13)$$

where

N_c, N_q, N_γ = bearing capacity factors [for N_c and N_q use Table 2.3; for N_γ see Table 2.4—equations (2.72), (2.73), (2.74)]
$\lambda_{cs}, \lambda_{qs}, \lambda_{\gamma s}$ = shape factors (Table 2.6)
$\lambda_{cd}, \lambda_{qd}, \lambda_{\gamma d}$ = depth factors (Table 2.6)
$\lambda_{ci}, \lambda_{qi}, \lambda_{\gamma i}$ = inclination factors

Meyerhof[4] provided the following inclination factor relationships:

$$\lambda_{ci} = \lambda_{qi} = \left(1-\frac{\alpha^\circ}{90^\circ}\right)^2 \quad (3.14)$$

$$\lambda_{\gamma i} = \left(1-\frac{\alpha^\circ}{\phi^\circ}\right)^2 \quad (3.15)$$

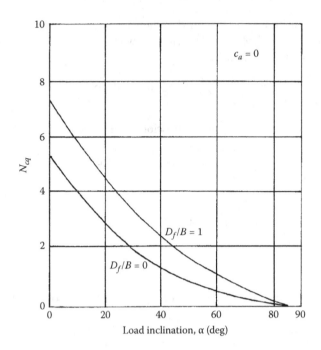

FIGURE 3.2 Meyerhof's bearing capacity factor N_{cq} for purely cohesive soil ($\phi = 0$). *Source:* Meyerhof, G. G. 1953. The bearing capacity of foundations under eccentric and inclined loads, in *Proc., III Intl. Conf. Soil Mech. Found. Eng.*, Zurich, Switzerland, 1: 440.

Hansen[5] also suggested the following relationships for inclination factors:

$$\lambda_{qi} = \left(1 - \frac{0.5 Q_u \sin\alpha}{Q_u \cos\alpha + BLc \cot\phi}\right)^5 \tag{3.16}$$

$$\lambda_{ci} = \lambda_{qi} - \left(\frac{1 - \lambda_{qi}}{\underset{\text{Table 2.3}}{N_c} - 1}\right) \tag{3.17}$$

$$\lambda_{\gamma i} = \left(1 - \frac{0.7 Q_u \sin\alpha}{Q_u \cos\alpha + BLc \cot\phi}\right)^5 \tag{3.18}$$

where, in equations (3.14) to (3.18),
α = inclination of the load on the foundation with the vertical
Q_u = ultimate load on the foundation = $q_u BL$
B = width of the foundation
L = length of the foundation

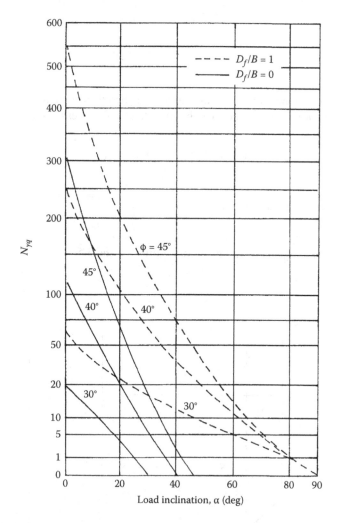

FIGURE 3.3 Meyerhof's bearing capacity factor $N_{\gamma q}$ for cohesionless soil ($\alpha = 0$, $\delta = \phi$). *Source:* Meyerhof, G. G. 1953. The bearing capacity of foundations under eccentric and inclined loads, in *Proc., III Intl. Conf. Soil Mech. Found. Eng.*, Zurich, Switzerland, 1: 440.

3.2.3 OTHER RESULTS FOR FOUNDATIONS WITH CENTRIC INCLINED LOAD

Based on the results of field tests, Muhs and Weiss[6] concluded that the ratio of the vertical component $Q_{u(v)}$ of the ultimate load with inclination α with the vertical to the ultimate load Q_u when the load is vertical (that is, $\alpha = 0$) is approximately equal to $(1 - \tan \alpha)^2$:

$$\frac{Q_{u(v)}}{Q_{u(\alpha=0)}} = (1 - \tan \alpha)^2$$

or

$$\frac{\frac{Q_{u(v)}}{BL}}{\frac{Q_{u(\alpha=0)}}{BL}} = \frac{q_{u(v)}}{q_{u(\alpha=0)}} = (1 - \tan\alpha)^2 \tag{3.19}$$

Dubrova[7] developed a theoretical solution for the ultimate bearing capacity of a *continuous foundation* with a centric inclined load and expressed it in the following form:

$$q_u = c(N_q^* - 1)\cot\phi + 2qN_q^* + B\gamma N_\gamma^* \tag{3.20}$$

where

$N_q^*, N_\gamma^* =$ bearing capacity factors
$q = \gamma D_f$

The variations of N_q^* and N_γ^* are given in Figures 3.4 and 3.5.

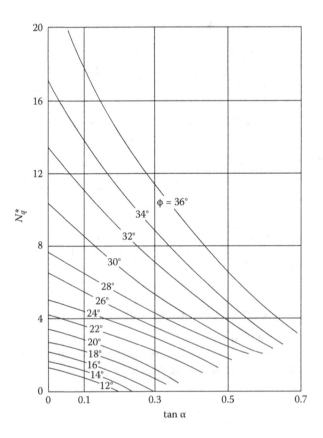

FIGURE 3.4 Variation of N_q^*.

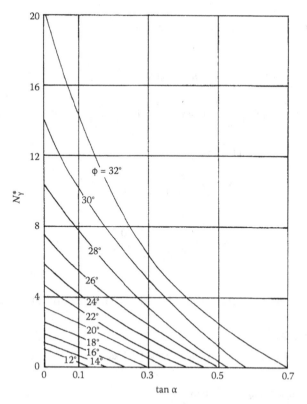

FIGURE 3.5 Variation of N_γ^*.

EXAMPLE 3.1

Consider a continuous foundation in a granular soil with the following: $B = 1.2$ m; $D_f = 1.2$ m; unit weight of soil $\gamma = 17$ kN/m³; soil friction angle $\phi = 40°$; load inclination $\alpha = 20°$. Calculate the gross ultimate load bearing capacity q_u.

   a. Use equation (3.12).
   b. Use equation (3.13) and Meyerhof's bearing capacity factors (Table 2.3), his shape and depth factors (Table 2.6), and inclination factors [equations (3.14) and (3.15)].

Solution

Part a
From equation (3.12),

$$q_u = \frac{\gamma B N_{\gamma q}}{2\cos\alpha}$$

where $\dfrac{D_f}{B} = \dfrac{1.2}{1.2} = 1$; $\phi = 40°$; and $\alpha = 20°$. From Figure 3.3, $N_{\gamma q} \approx 100$. So,

$$qu = \frac{(17)(1.2)(100)}{2\cos 20} = 1085.5 \text{ kN/m}^2$$

Part b

With $c = 0$ and $B/L = 0$, equation (3.13) becomes

$$q_u = qN_q\lambda_{qd}\lambda_{qi} + \tfrac{1}{2}\gamma BN_\gamma\lambda_{\gamma d}\lambda_{\gamma i}$$

For $\phi = 40°$, from Table 2.3, $N_q = 64.2$ and $N_\gamma = 93.69$. From Table 2.6,

$$\lambda_{qd} = \lambda_{\gamma d} = 1 + 0.1\left(\frac{D_f}{B}\right)\tan\left(45 + \frac{\phi}{2}\right) = 1 + 0.1\left(\frac{1.2}{1.2}\right)\tan\left(45 + \frac{40}{2}\right) = 1.214$$

From equations (3.14) and (3.15),

$$\lambda_{qi} = \left(1 - \frac{\alpha°}{90°}\right)^2 = \left(1 - \frac{20}{90}\right)^2 = 0.605$$

$$\lambda_{\gamma i} = \left(1 - \frac{\alpha°}{\phi°}\right)^2 = \left(1 - \frac{20}{40}\right)^2 = 0.25.$$

So,

$$q_u = (1.2\times17)(64.2)(1.214)(0.605) + \tfrac{1}{2}(17)(1.2)(93.69)(1.214)(0.25) = 1252 \text{ kN/m}^2$$

EXAMPLE 3.2

Consider the continuous foundation described in Example 3.1. Other quantities remaining the same, let $\phi = 35°$.

  a. Calculate q_u using equation (3.12).
  b. Calculate q_u using equation (3.20).

Solution

Part a

From equation (3.12),

$$q_u = \frac{\gamma BN_{\gamma q}}{2\cos\alpha}$$

From Figure 3.3, $N_{\gamma q} \approx 65$:

$$q_u = \frac{(17)(1.2)(65)}{2\cos 20} \approx 706 \text{ kN/m}^2$$

Part b

For $c = 0$, equation (3.20) becomes

$$q_u = 2qN_q^* + B\gamma N_\gamma^*$$

Using Figures 3.4 and 3.5 for $\phi = 35°$ and $\tan\alpha = \tan 20 = 0.36$, $N_q^* \approx 8.5$ and $N_\gamma^* \approx 6.5$ (extrapolation):

$$q_u = (2)(17\times1.2)(8.5) + (1.2)(17)(6.5) \approx 480 \text{ kN/m}^2$$

Note: Equation (3.20) does not provide depth factors.

3.3 FOUNDATIONS SUBJECTED TO ECCENTRIC LOAD

3.3.1 CONTINUOUS FOUNDATION WITH ECCENTRIC LOAD

When a shallow foundation is subjected to an eccentric load, it is assumed that the contact pressure decreases linearly from the toe to the heel; however, at ultimate load, the contact pressure is not linear. This problem was analyzed by Meyerhof[1] who suggested the concept of *effective width B'*. The effective width is defined as (Figure 3.6)

$$B' = B - 2e \qquad (3.21)$$

where

e = load eccentricity

According to this concept, the bearing capacity of a continuous foundation can be determined by assuming that the load acts centrally along the effective contact width as shown in Figure 3.6. Thus, for a continuous foundation [from equation (2.82)] with vertical loading,

$$q_u = cN_c\lambda_{cd} + qN_q\lambda_{qd} + \tfrac{1}{2}\gamma B'N_\gamma\lambda_{\gamma d} \qquad (3.22)$$

Note that the shape factors for a continuous foundation are equal to one. The ultimate load *per unit length* of the foundation Q_u can now be calculated as

$$Q_u = q_u A'$$

where

A' = effective area = $B' \times 1 = B'$

3.3.1.1 Reduction Factor Method

Purkayastha and Char[8] carried out stability analyses of eccentrically loaded *continuous foundations* supported by sand ($c = 0$) using the method of slices proposed by Janbu.[9] Based on that analysis, they proposed that

$$R_k = 1 - \frac{q_{u(\text{eccentric})}}{q_{u(\text{centric})}} \qquad (3.23)$$

where

R_k = reduction factor

$q_{u(\text{eccentric})}$ = ultimate bearing capacity of eccentrically loaded continuous foundations

$q_{u(\text{centric})}$ = ultimate bearing capacity of centrally loaded continuous foundations

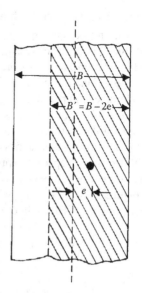

FIGURE 3.6 Effective width B'.

TABLE 3.1
Variations of a and k [Equation (3.24)]

D_f/B	a	k
0.00	1.862	0.73
0.25	1.811	0.785
0.50	1.754	0.80
1.00	1.820	0.888

The magnitude of R_k can be expressed as

$$R_k = a\left(\frac{e}{B}\right)^k \tag{3.24}$$

where a and k are functions of the embedment ratio D_f/B (Table 3.1).

Hence, combining equations (3.23) and (3.24),

$$q_{u(\text{eccentric})} = q_{u(\text{centric})}(1 - R_k) = q_{u(\text{centric})}\left[1 - a\left(\frac{e}{B}\right)^k\right] \tag{3.25}$$

where

$$q_{u(\text{centric})} = qN_q\lambda_{qd} + \tfrac{1}{2}\gamma BN_\gamma\lambda_{\gamma d} \quad (\text{Note: } c = 0) \tag{3.26}$$

3.3.1.2 Theory of Prakash and Saran

Prakash and Saran[10] provided a comprehensive mathematical formulation to estimate the ultimate bearing capacity for rough *continuous foundations* under eccentric loading. According to this procedure, Figure 3.7 shows the assumed failure surface in a $c–\phi$ soil under a continuous foundation subjected to eccentric loading. Let Q_u be the ultimate load per unit length of the foundation of width B with an eccentricity e. In Figure 3.7 zone I is an elastic zone with wedge angles of ψ_1 and ψ_2. Zones II and III are similar to those assumed by Terzaghi (that is, zone II is a radial shear zone and zone III is a Rankine passive zone).

The bearing capacity expression can be developed by considering the equilibrium of the elastic wedge *abc* located below the foundation (Figure 3.7b). Note that in Figure 3.7b the contact width of the foundation with the soil is equal to Bx_1. Neglecting the self-weight of the wedge,

$$Q_u = P_p\cos(\psi_1 - \phi) + P_m\cos(\psi_2 - \phi_m) + C_a\sin\psi_1 + C_a'\sin\psi_2 \tag{3.27}$$

where
   P_p, P_m = passive forces per unit length of the wedge along the wedge faces *bc* and
      ac, respectively

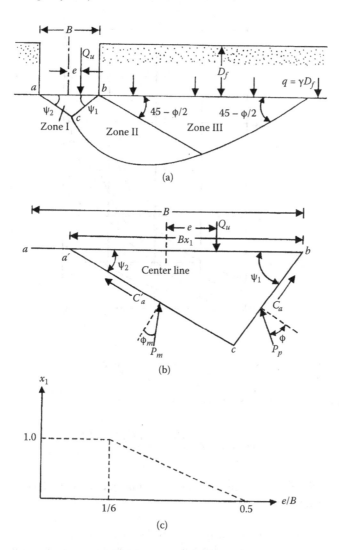

FIGURE 3.7 Derivation of the bearing capacity theory of Prakash and Saran for eccentrically loaded rough continuous foundation.

ϕ = soil friction angle

ϕ_m = mobilized soil friction angle ($\leq \phi$)

C_a = adhesion along wedge face $bc = \dfrac{cBx_1 \sin \psi_2}{\sin(\psi_1 + \psi_2)}$

C'_a = adhesion along wedge face $ac = \dfrac{mcBx_1 \sin \psi_1}{\sin(\psi_1 + \psi_2)}$

m = mobilization factor (≤ 1)

c = unit cohesion

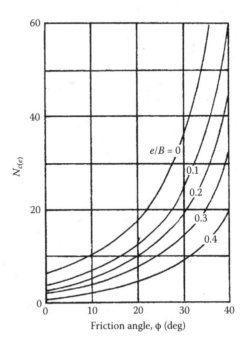

FIGURE 3.8 Prakash and Saran's bearing capacity factor $N_{c(e)}$.

Equation (3.27) can be expressed in the form

$$q_u = \frac{Q_u}{(B \times 1)} = \frac{1}{2}\gamma B N_{\gamma(e)} + \gamma D_f N_{q(e)} + c N_{c(e)} \tag{3.28}$$

where

$N_{\gamma(e)}, N_{q(e)}, N_{c(e)}$ = bearing capacity factors for an eccentrically loaded continuous foundation

The above-stated bearing capacity factors will be functions of e/B, ϕ, and also the foundation contact factor x_1. In obtaining the bearing capacity factors, Prakash and Saran[10] assumed the variation of x_1 as shown in Figure 3.7c. Figures 3.8, 3.9, and 3.10 show the variations of $N_{\gamma(e)}, N_{q(e)}$, and $N_{c(e)}$ with ϕ and e/B. Note that, for $e/B = 0$, the bearing capacity factors coincide with those given by Terzaghi[11] for a centrically loaded foundation.

Prakash[12] also gave the relationships for the settlement of a given foundation under centric and eccentric loading conditions for an equal factor of safety *FS*. They are as follows (Figure 3.11):

$$\frac{S_e}{S_o} = 1.0 - 1.63\left(\frac{e}{B}\right) - 2.63\left(\frac{e}{B}\right)^2 + 5.83\left(\frac{e}{B}\right)^3 \tag{3.29}$$

$$\frac{S_m}{S_o} = 1.0 - 2.31\left(\frac{e}{B}\right) - 22.61\left(\frac{e}{B}\right)^2 + 31.54\left(\frac{e}{B}\right)^3 \tag{3.30}$$

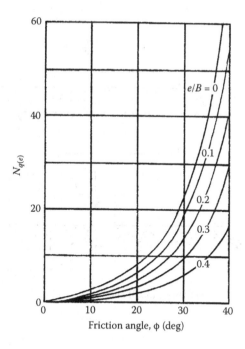

FIGURE 3.9 Prakash and Saran's bearing capacity factor $N_{q(e)}$.

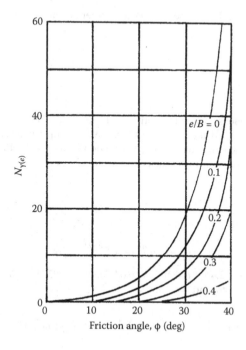

FIGURE 3.10 Prakash and Saran's bearing capacity factor $N_{\gamma(e)}$.

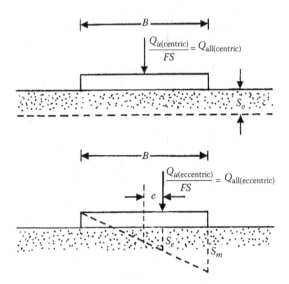

FIGURE 3.11 Notations for equations (3.28) and (3.29).

where

S_o = settlement of a foundation under centric loading at $q_{all(centric)} = \dfrac{q_{u(centric)}}{FS}$

S_e, S_m = settlements of the same foundation under eccentric loading at $q_{all(eccentric)}$

$= \dfrac{q_{u(eccentric)}}{FS}$

EXAMPLE 3.3

Consider a continuous foundation with a width of 2 m. If $e = 0.2$ m and the depth of the foundation $D_f = 1$ m, determine the ultimate load per unit meter length of the foundation using the reduction factor method. For the soil, use $\phi = 40°$; $\gamma = 17.5$ kN/m³; $c = 0$. Use Meyerhof's bearing capacity and depth factors.

Solution

Since $c = 0$, $B/L = 0$. From equation (3.26),

$$q_{u(centric)} = qN_q\lambda_{qd} + \tfrac{1}{2}\gamma BN_\gamma\lambda_{\gamma d}$$

From Tables 2.3 and 2.4 for $\phi = 40°$, $N_q = 64.2$ and $N_\gamma = 93.69$. Again, from Table 2.6, Meyerhof's depth factors are as follows:

$$\lambda_{qd} = \lambda_{\gamma d} = 1 + 0.1\left(\frac{D_f}{B}\right)\tan\left(45 + \frac{\phi}{2}\right) = 1 + 0.1\left(\frac{1}{2}\right)\tan\left(45 + \frac{40}{2}\right) = 1.107$$

So,

$$q_{u(centric)} = (1)(17.5)(64.2)(1.107) + \tfrac{1}{2}(17.5)(2)(93.69)(1.107) = 1243.7 + 1815.0 = 3058.7 \text{ kN/m}^2$$

According to equation (3.25),

$$q_{u(eccentric)} = q_{u(centric)}(1 - R_k) = q_{u(centric)}\left[1 - a\left(\frac{e}{B}\right)^k\right]$$

For $D_f/B = 1/2 = 0.5$, from Table 3.1 $a = 1.754$ and $k = 0.80$. So,

$$q_{u(eccentric)} = 3058.7\left[1 - 1.754\left(\frac{0.2}{2}\right)^{0.8}\right] \approx 2209 \text{ kN/m}^2$$

The ultimate load per unit length:

$$Q = (2209)(B)(1) = (2209)(2)(1) = \mathbf{4418 \text{ kN/m}}$$

Example 3.4

Solve the Example 3.3 problem using the method of Prakash and Saran.

Solution

From equation (3.28),

$$Q_u = (B \times 1)\left[\tfrac{1}{2}\gamma BN_{\gamma(e)} + \gamma D_f N_{q(e)} + cN_{c(e)}\right]$$

Given: $c = 0$. For $\phi = 40°$, $e/B = 0.2/2 = 0.1$. From Figures 3.9 and 3.10, $N_{q(e)} = 56.09$ and $N_{\gamma(e)} \approx 55$. So,

$$Q_u = (2 \times 1)\left[\tfrac{1}{2}(17.5)(2)(55) + (17.5)(1)(56.09)\right] = (2)(962.5 + 981.5) = \mathbf{3888 \text{ kN/m}}$$

Example 3.5

Solve the Example 3.3 problem using equation (3.22).

Solution

For $c = 0$, from equation (3.22),

$$q_u = qN_q\lambda_{qd} + \tfrac{1}{2}\gamma B'N_\gamma\lambda_{\gamma d}$$

$$B' = B - 2e = 2 - (2)(0.2) = 1.6 \text{ m}$$

From Tables 2.3 and 2.4, $N_q = 64.2$ and $N_\gamma = 93.69$. From Table 2.6, Meyerhof's depth factors are as follows:

$$\lambda_{qd} = \lambda_{\gamma d} = 1 + 0.1 \left(\frac{D_f}{B} \right) \tan \left(45 + \frac{\phi}{2} \right) = 1 + 0.1 \left(\frac{1}{2} \right) \tan \left(45 + \frac{40}{2} \right) = 1.107$$

$$q_u = (1 \times 17.5)(64.2)(1.107) + \tfrac{1}{2}(17.5)(1.6)(93.69)(1.107) = 2695.9 \text{ kN/m}^2$$

$$Q_u = (B' \times 1)q_u = (1.6)(2695.5) \approx \mathbf{4313 \text{ kN/m}}$$

3.3.2 ULTIMATE LOAD ON RECTANGULAR FOUNDATION

Meyerhof's effective area method[1] described in the preceding section can be extended to determine the ultimate load on rectangular foundations. Eccentric loading of shallow foundations occurs when a vertical load Q is applied at a location other than the centroid of the foundation (Figure 3.12a), or when a foundation is subjected to a centric load of magnitude Q and momentum M (Figure 3.12b). In such cases, the load

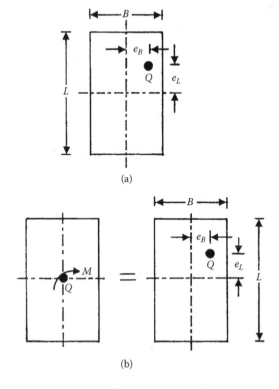

(a)

(b)

FIGURE 3.12 Eccentric load on rectangular foundation.

eccentricities may be given as

$$e_L = \frac{M_B}{Q} \qquad (3.31)$$

and

$$e_B = \frac{M_L}{Q} \qquad (3.32)$$

where

e_L, e_B = load eccentricities, respectively, in the directions of the *long* and *short* axes of the foundation

M_B, M_L = moment components about the *short* and *long* axes of the foundation, respectively

According to Meyerhof,[1] the ultimate bearing capacity q_u and the ultimate load Q_u of an eccentrically loaded foundation (vertical load) can be given as

$$q_u = cN_c \lambda_{cs} \lambda_{cd} + qN_q \lambda_{qs} \lambda_{qd} + \tfrac{1}{2} \gamma B' N_\gamma \lambda_{\gamma s} \lambda_{\gamma d} \qquad (3.33)$$

and

$$Q_u = (q_u)A' \qquad (3.34)$$

where

A' = effective area = $B'L'$

B' = effective width

L' = effective length

The *effective area* A' is a minimum contact area of the foundation such that its *centroid coincides with that of the load*. For *one-way eccentricity* [that is, if $e_L = 0$ (Figure 3.13a)],

$$B' = B - 2e_B; \; L' = L; \; A' = B'L \qquad (3.35)$$

However, if $e_B = 0$ (Figure 3.13b), calculate $L - 2e_L$. The effective area is

$$A' = B(L - 2e_L) \qquad (3.36)$$

The effective width B' is the *smaller* of the two values, that is, B or $L - 2e_L$.

Based on their model test results Prakash and Saran[10] suggested that, for rectangular foundations with *one-way eccentricity* in the width direction (Figure 3.14), the ultimate load may be expressed as

$$Q_u = q_u(BL) = (BL)\left[\tfrac{1}{2}\gamma BN_{\gamma(e)}\lambda_{\gamma s(e)} + \gamma D_f N_{q(e)}\lambda_{qs(e)} + cN_{c(e)}\lambda_{cs(e)}\right] \qquad (3.37)$$

where

$\lambda_{\gamma s(e)}, \lambda_{qs(e)}, \lambda_{cs(e)}$ = shape factors

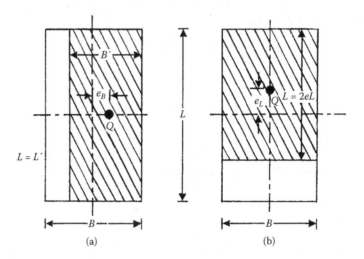

FIGURE 3.13 One-way eccentricity of load on foundation.

The shape factors may be expressed by the following relationships:

$$\lambda_{\gamma s(e)} = 1.0 + \left(\frac{2e_B}{B} - 0.68\right)\frac{B}{L} + \left(0.43 - \frac{3e_B}{2B}\right)\left(\frac{B}{L}\right)^2 \qquad (3.38)$$

where
　　　L = length of the foundation
　　$\lambda_{qs(e)} = 1$ $\qquad\qquad\qquad\qquad\qquad\qquad\qquad\qquad\qquad\qquad$ (3.39)

$\lambda_{cs(e)} = 1.0 + \left(\frac{B}{L}\right)$ $\qquad\qquad\qquad\qquad\qquad\qquad\qquad\qquad$ (3.40)

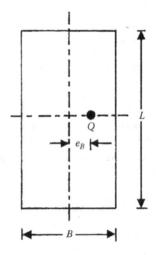

FIGURE 3.14 Rectangular foundation with one-way eccentricity.

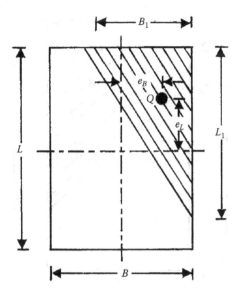

FIGURE 3.15 Effective area for the case of $e_L/L \geq 1/6$ and $e_B/B \geq 1/6$.

Note that equation (3.37) does not contain the depth factors.

For *two-way eccentricities* (that is, $e_L \neq 0$ and $e_B \neq 0$), five possible cases may arise as discussed by Highter and Anders.[13] They are as follows:

Case I ($e_L/L \geq 1/6$ and $e_B/B \geq 1/6$)

For this case (shown in Figure 3.15), calculate

$$B_1 = B\left(1.5 - \frac{3e_B}{B}\right) \tag{3.41}$$

$$L_1 = L\left(1.5 - \frac{3e_L}{L}\right) \tag{3.42}$$

So, the effective area

$$A' = \tfrac{1}{2}B_1 L_1 \tag{3.43}$$

The effective width B' is equal to the smaller of B_1 or L_1.

Case II ($e_L/L < 0.5$ and $0 < e_B/B < 1/6$)

This case is shown in Figure 3.16. Knowing the magnitudes e_L/L and e_B/B, the values of L_1/L and L_2/L (and thus L_1 and L_2) can be obtained from Figures 3.17 and 3.18. The effective area is given as

$$A' = \tfrac{1}{2}(L_1 + L_2)B \tag{3.44}$$

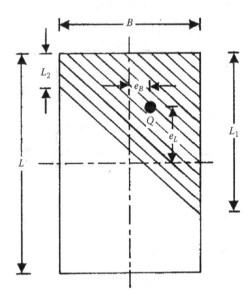

FIGURE 3.16 Effective area for the case of $e_L/L < 0.5$ and $e_B/B < 1/6$.

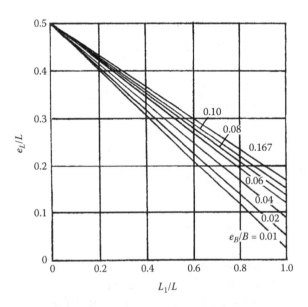

FIGURE 3.17 Plot of e_L/L versus L_1/L for $e_L/L < 0.5$ and $0 < e_B/B < 1/6$. *Source:* Redrawn from Highter, W. H., and J. C. Anders. 1985. Dimensioning footings subjected to eccentric loads. *J. Geotech. Eng.*, ASCE, 111(5): 659.

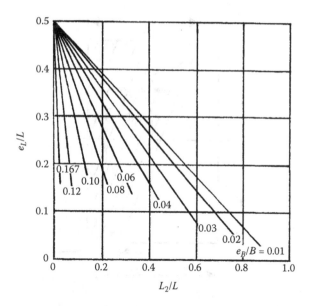

FIGURE 3.18 Plot of e_L/L versus L_2/L for $e_L/L < 0.5$ and $0 < e_B/B < 1/6$. *Source:* Redrawn from Highter, W. H., and J. C. Anders. 1985. Dimensioning footings subjected to eccentric loads. *J. Geotech. Eng.*, ASCE, 111(5): 659.

The effective length L' is the larger of the two values L_1 or L_2. The effective width is equal to

$$B' = \frac{A'}{L'} \qquad (3.45)$$

Case III ($e_L/L < 1/6$ and $0 < e_B/B < 0.5$)

Figure 3.19 shows the case under consideration. Knowing the magnitudes of e_L/L and e_B/B, the magnitudes of B_1 and B_2 can be obtained from Figures 3.20 and 3.21. So, the effective area can be obtained as

$$A' = \tfrac{1}{2}(B_1 + B_2)L \qquad (3.46)$$

In this case, the effective length is equal to

$$L' = L \qquad (3.47)$$

The effective width can be given as

$$B' = \frac{A'}{L} \qquad (3.48)$$

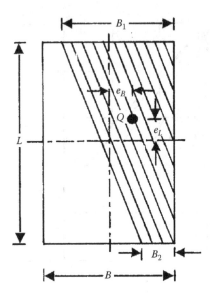

FIGURE 3.19 Effective area for the case of $e_L/L < 1/6$ and $0 < e_B/B < 0.5$.

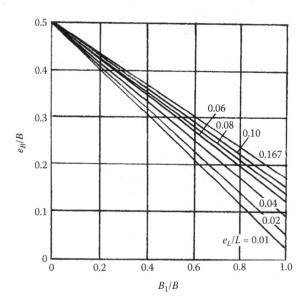

FIGURE 3.20 Plot of e_B/B versus B_1/B for $e_L/L < 1/6$ and $0 < e_B/B < 0.5$. *Source:* Redrawn from Highter, W. H., and J. C. Anders. 1985. Dimensioning footings subjected to eccentric loads. *J. Geotech. Eng.*, ASCE, 111(5): 659.

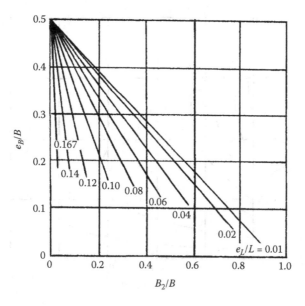

FIGURE 3.21 Plot of e_B/B versus B_2/B for $e_L/L < 1/6$ and $0 < e_B/B < 0.5$. *Source:* Redrawn from Highter, W. H., and J. C. Anders. 1985. Dimensioning footings subjected to eccentric loads. *J. Geotech. Eng.*, ASCE, 111(5): 659.

Case IV ($e_L/L < 1/6$ and $e_B/B < 1/6$)

The eccentrically loaded plan of the foundation for this condition is shown in Figure 3.22. For this case, the e_L/L curves sloping upward in Figure 3.23 represent the values of B_2/B on the abscissa. Similarly, in Figure 3.24 the families of e_L/L

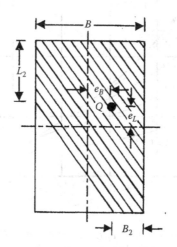

FIGURE 3.22 Effective area for the case of $e_L/L < 1/6$ and $e_B/B < 1/6$.

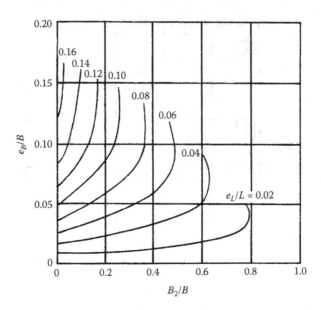

FIGURE 3.23 Plot of e_B/B versus B_2/B for $e_L/L < 1/6$ and $e_B/B < 1/6$. *Source:* Redrawn from Highter, W. H., and J. C. Anders. 1985. Dimensioning footings subjected to eccentric loads. *J. Geotech. Eng.*, ASCE, 111(5):659.

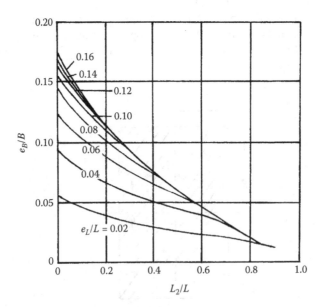

FIGURE 3.24 Plot of e_B/B versus L_2/L for $e_L/L < 1/6$ and $e_B/B < 1/6$. *Source:* Redrawn from Highter, W. H., and J. C. Anders. 1985. Dimensioning footings subjected to eccentric loads. *J. Geotech. Eng.*, ASCE, 111(5): 659.

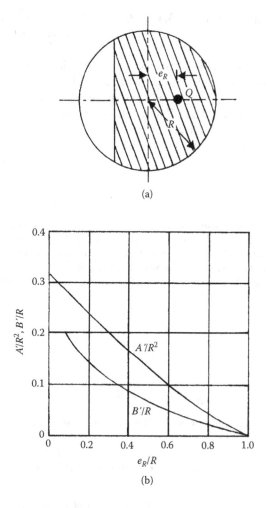

FIGURE 3.25 Normalized effective dimensions of circular foundations. *Source:* From Highter, W. H., and J. C. Anders. 1985. Dimensioning footings subjected to eccentric loads. *J. Geotech. Eng.*, ASCE, 111(5): 659.

curves that slope downward represent the values of L_2/L on the abscissa. Knowing B_2 and L_2, the effective area A' can be calculated. For this case, $L' = L$ and $B' = A'/L'$.

Case V (Circular Foundation)

In the case of circular foundations under eccentric loading (Figure 3.25a), the eccentricity is always one way. The effective area A' and the effective width B' for a circular foundation are given in a nondimensional form in Figure 3.25b.

Depending on the nature of the load eccentricity and the shape of the foundation, once the magnitudes of the effective area and the effective width are determined,

102 Shallow Foundations: Bearing Capacity and Settlement

they can be used in equations (3.33) and (3.34) to determine the ultimate load for the foundation. In using equation (3.33), one needs to remember that

1. The bearing capacity factors for a given friction angle are to be determined from those presented in Tables 2.3 and 2.4.
2. The shape factor is determined by using the relationships given in Table 2.6 by replacing B' for B and L' for L whenever they appear.
3. The depth factors are determined from the relationships given in Table 2.6. However, for calculating the depth factor, the term B is *not* replaced by B'.

EXAMPLE 3.5

A shallow foundation measuring 2 m $\times$ 3 m in a plan is subjected to a centric load and a moment. If $e_B = 0.2$ m, $e_L = 0.6$ m, and the depth of the foundation is 1.5 m, determine the allowable load the foundation can carry. Use a factor of safety of 4. For the soil, given: unit weight $\gamma = 18$ kN/m³; friction angle $\phi = 35°$; cohesion $c = 0$. Use Vesic's N_γ (Table 2.4), DeBeer's shape factors (Table 2.6), and Hansen's depth factors (Table 2.6).

Solution

For this case,

$$\frac{e_B}{B} = \frac{0.2}{2} = 0.1; \quad \frac{e_L}{L} = \frac{0.6}{3} = 0.2.$$

For this type of condition, Case II as shown in Figure 3.16 applies. Referring to Figures 3.17 and 3.18,

$$\frac{L_1}{L} = 0.865, \text{ or } L_1 = (0.865)(3) = 2.595 \text{ m}$$

$$\frac{L_2}{L} = 0.22, \text{ or } L_2 = (0.22)(3) = 0.66 \text{ m}$$

From equation (3.44),

$$A' = \tfrac{1}{2}(L_1 + L_2)B = \tfrac{1}{2}(2.595 + 0.66)(2) = 3.255 \text{ m}^2$$

So,

$$B' = \frac{A'}{L'} = \frac{A'}{L_1} = \frac{3.255}{2.595} = 1.254 \text{ m}$$

Since $c = 0$,

$$q_u = qN_q\lambda_{qs}\lambda_{qd} + \tfrac{1}{2}\gamma B'N_\gamma\lambda_{\gamma s}\lambda_{\gamma d}$$

From Table 2.3 for $\phi = 35°$, $N_q = 33.30$. Also from Table 2.4 for $\phi = 35°$, Vesic's $N_\gamma = 48.03$.

The shape factors given by DeBeer are as follows (Table 2.6):

$$\lambda_{qs} = 1 + \left(\frac{B'}{L'}\right)\tan\phi = 1 + \left(\frac{1.254}{2.595}\right)\tan 35 = 1.339$$

$$\lambda_{\gamma s} = 1 - 0.4\left(\frac{B'}{L'}\right) = 1 - 0.4\left(\frac{1.254}{2.595}\right) = 0.806$$

The depth factors given by Hansen are as follows:

$$\lambda_{qd} = 1 + 2\tan\phi(1-\sin\phi)^2\left(\frac{D_f}{B}\right) = 1 + (2)(\tan 35)(1-\sin 35)^2\left(\frac{1.5}{2}\right) = 1.191$$

$$\lambda_{\gamma d} = 1$$

So,

$$q_u = (18)(1.5)(33.3)(1.339)(1.191) + \tfrac{1}{2}(18)(1.254)(48.03)(0.806)(1) = 1434 + 437 = 1871 \text{ kN/m}^2$$

So the allowable load on the foundation is

$$Q = \frac{qA'}{FS} = \frac{(1871)(3.255)}{4} \approx \mathbf{1523 \text{ kN}}$$

3.3.3 ULTIMATE BEARING CAPACITY OF ECCENTRICALLY OBLIQUELY LOADED FOUNDATIONS

The problem of ultimate bearing capacity of a *continuous foundation* subjected to an eccentric inclined load was studied by Saran and Agarwal.[14] If a continuous foundation is located at a depth D_f below the ground surface and is subjected to an eccentric load (load eccentricity = e) inclined at an angle α to the vertical, the ultimate capacity can be expressed as

$$Q_{ult} = B\left[cN_{c(ei)} + qN_{q(ei)} + \tfrac{1}{2}\gamma BN_{\gamma(ei)}\right] \tag{3.49}$$

where
$N_{c(ei)}, N_{q(ei)}, N_{\gamma(ei)}$ = bearing capacity factors
$q = \gamma D_f$

The variations of the bearing capacity factors with e/B, ϕ, and α are given in Figures 3.26, 3.27, and 3.28.

EXAMPLE 3.6

For a continuous foundation, given: $B = 1.5$ m; $D_f = 1$ m; $\gamma = 16$ kN/m^3; eccentricity $e = 0.15$ m; load inclination $\alpha = 20°$. Estimate the ultimate load Q_{ult}.

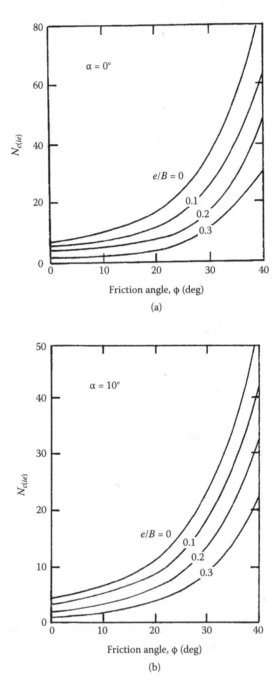

FIGURE 3.26 Variation of $N_{c(ie)}$ with soil friction angle ϕ and e/B.

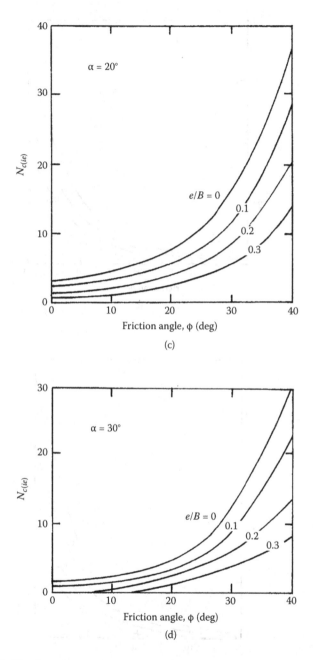

FIGURE 3.26 (Continued).

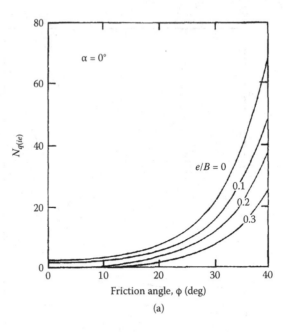

(a)

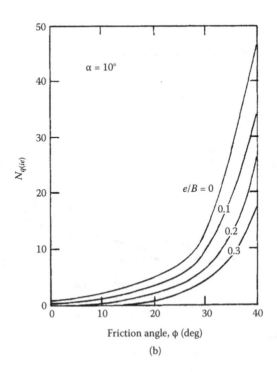

(b)

FIGURE 3.27 Variation of $N_{q(ie)}$ with soil friction angle ϕ and e/B.

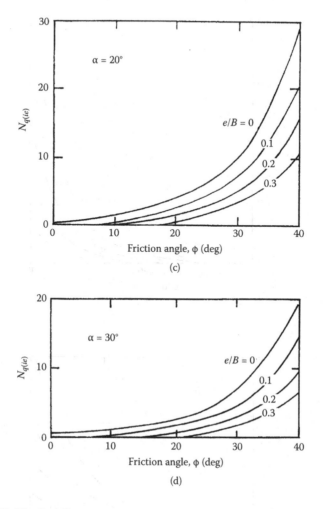

FIGURE 3.27 (Continued).

Solution

With $c = 0$, from equation (3.49),

$$Q_{ult} = B\left[qN_{q(ei)} + \tfrac{1}{2}\gamma BN_{\gamma(ei)}\right]$$

$B = 1.5$ m, $q = \gamma D_f = (1)(16) = 16$ kN/m², $e/B = 0.15/1.5 = 0.1$, and $\alpha = 20°$. From Figures 3.27c and 3.28c, $N_{q(ei)} = 14.2$ and $N_{\gamma(ei)} = 20$. Hence,

$$Q_{ult} = (1.5)\left[(16)(14.2) + \tfrac{1}{2}(16)(1.5)(20)\right] = \mathbf{700.8\ kN/m}$$

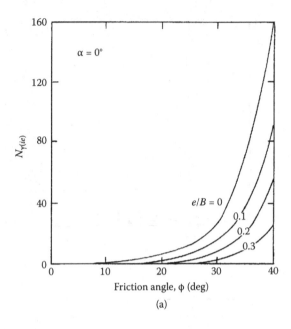

(a)

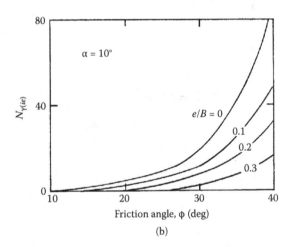

(b)

FIGURE 3.28 Variation of $N_{\gamma(ie)}$ with soil friction angle ϕ and e/B.

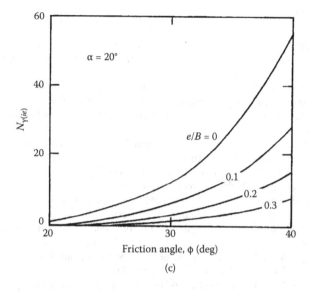

(c)

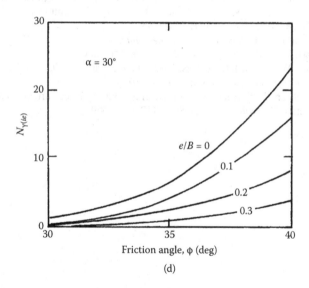

(d)

FIGURE 3.28 (Continued).

REFERENCES

1. Meyerhof, G. G. 1953. The bearing capacity of foundations under eccentric and inclined loads, in *Proc., III Intl. Conf. Soil Mech. Found. Eng.*, Zurich, Switzerland, 1: 440.
2. Caquot, A., and J. Kerisel. 1949. *Tables for the calculation of passive pressure, active pressure, and the bearing capacity of foundations.* Paris: Gauthier–Villars.
3. Meyerhof, G. G. 1951. The ultimate bearing capacity of foundations. *Geotechnique.* 2: 301.
4. Meyerhof, G. G. 1963. Some recent research on the bearing capacity of foundations. *Canadian Geotech. J.* 1(1): 16.
5. Hansen, J. B. 1970. *A revised and extended formula for bearing capacity.* Bulletin No. 28. Copenhagen: Danish Geotechnical Institute.
6. Muhs, H., and K. Weiss. 1973. Inclined load tests on shallow strip footing, in *Proc., VIII Int. Conf. Soil Mech. Found. Eng.*, Moscow, 1:3.
7. Dubrova, G. A. 1973. *Interaction of soils and structures.* Moscow: Rechnoy Transport.
8. Purkayastha, R. D., and R. A. N. Char. 1977. Stability analysis for eccentrically loaded footings. *J. Geotech. Eng. Div.*, ASCE, 103(6): 647.
9. Janbu, N. 1957. Earth pressures and bearing capacity calculations by generalized procedure of slices, in *Proc., IV Int. Conf. Soil Mech. Found. Eng.*, London, 2: 207.
10. Prakash, S., and S. Saran. 1971. Bearing capacity of eccentrically loaded footings. *J. Soil Mech. Found. Div.*, ASCE, 97(1): 95.
11. Terzaghi, K. 1943. *Theoretical soil mechanics.* New York: John Wiley.
12. Prakash, S. 1981. *Soil dynamics.* New York: McGraw-Hill.
13. Highter, W. H., and J. C. Anders. 1985. Dimensioning footings subjected to eccentric loads. *J. Geotech. Eng.*, ASCE, 111(5): 659.
14. Saran, S., and R. K. Agarwal. 1991. Bearing capacity of eccentrically obliquely loaded foundation. *J. Geotech. Eng.*, ASCE, 117(11): 1669.

4 Special Cases of Shallow Foundations

4.1 INTRODUCTION

The bearing capacity problems described in Chapters 2 and 3 assume that the soil supporting the foundation is homogeneous and extends to a great depth below the bottom of the foundation. They also assume that the ground surface is horizontal; however, this is not true in all cases. It is possible to encounter a rigid layer at a shallow depth, or the soil may be layered and have different shear strength parameters. It may be necessary to construct foundations on or near a slope. Bearing capacity problems related to these special cases are described in this chapter.

4.2 FOUNDATION SUPPORTED BY SOIL WITH A RIGID ROUGH BASE AT A LIMITED DEPTH

Figure 4.1a shows a shallow rigid rough continuous foundation supported by soil that extends to a great depth. The ultimate bearing capacity of this foundation can be expressed (neglecting the depth factors) as (Chapter 2)

$$q_u = cN_c + qN_q + \tfrac{1}{2}\gamma BN_\gamma \qquad (4.1)$$

The procedure for determining the bearing capacity factors N_c, N_q, and N_γ in homogeneous and isotropic soils was outlined in Chapter 2. The extent of the failure zone in soil at ultimate load q_u is equal to D. The magnitude of D obtained during the evaluation of the bearing capacity factor N_c by Prandtl[1] and N_q by Reissner[2] is given in a nondimensional form in Figure 4.2. Similarly, the magnitude of D obtained by Lundgren and Mortensen[3] during the evaluation of N_γ is given in Figure 4.3.

If a rigid rough base is located at a depth of $H < D$ below the bottom of the foundation, full development of the failure surface in soil will be restricted. In such a case, the soil failure zone and the development of slip lines at ultimate load will be as shown in Figure 4.1b. Mandel and Salencon[4] determined the bearing capacity factors for such a case by numerical integration using the theory of plasticity. According to Mandel and Salencon's theory, the ultimate bearing capacity of a rough continuous foundation with a rigid rough base located at a shallow depth can be given by the relation

$$q_u = cN_c^* + qN_q^* + \tfrac{1}{2}\gamma BN_\gamma^* \qquad (4.2)$$

111

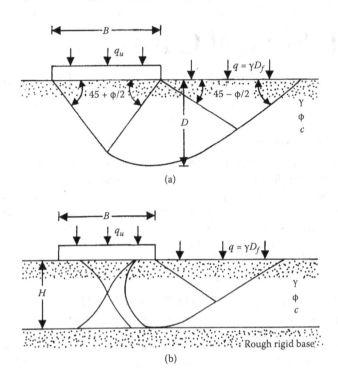

FIGURE 4.1 Failure surface under a rigid rough continuous foundation: (a) homogeneous soil extending to a great depth; (b) with a rough rigid base located at a shallow depth.

where

N_c^*, N_q^*, N_γ^* = modified bearing capacity factors

B = width of foundation

γ = unit weight of soil

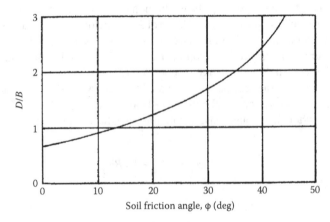

FIGURE 4.2 Variation of D/B with soil friction angle (for N_c and N_q).

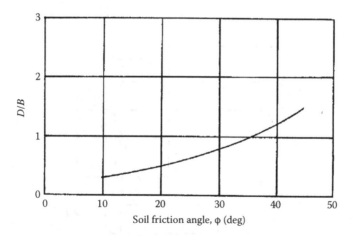

FIGURE 4.3 Variation of D/B with soil friction angle (for N_γ).

Note that for $H \geq D$, $N_c^* = N_c$, $N_q^* = N_q$, and $N_\gamma^* = N_\gamma$ (Lundgren and Mortensen). The variations of N_c N_q^*, and N_γ^* with H/B and soil friction angle ϕ are given in Figures 4.4, 4.5, and 4.6, respectively.

Neglecting the depth factors, the ultimate bearing capacity of rough circular and rectangular foundations on a sand layer ($c = 0$) with a rough rigid base located at a shallow depth can be given as

$$q_u = qN_q^* \lambda_{qs}^* + \tfrac{1}{2} \gamma BN_\gamma^* \lambda_{\gamma s}^* \tag{4.3}$$

where
λ_{qs}^*, $\lambda_{\gamma s}^*$ = modified shape factors

The above-mentioned shape factors are functions of H/B and ϕ. Based on the work of Meyerhof and Chaplin[5] and simplifying the assumption that the stresses and shear zones in radial planes are identical to those in transverse planes, Meyerhof[6] evaluated the approximate values of λ_{qs}^* and $\lambda_{\gamma s}^*$ as

$$\lambda_{qs}^* = 1 - m_1 \left(\frac{B}{L} \right) \tag{4.4}$$

and

$$\lambda_{\gamma s}^* = 1 - m_2 \left(\frac{B}{L} \right) \tag{4.5}$$

where
L = length of the foundation

The variations of m_1 and m_2 with H/B and ϕ are given in Figures 4.7 and 4.8.

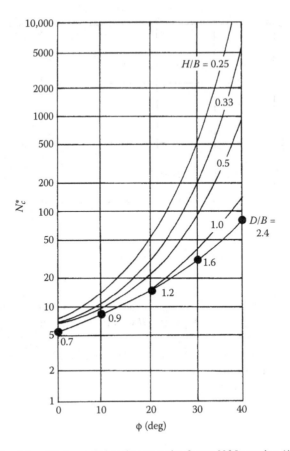

FIGURE 4.4 Mandel and Salencon's bearing capacity factor N_c^* [equation (4.2)].

Milovic and Tournier[7] and Pfeifle and Das[8] conducted laboratory tests to verify the theory of Mandel and Salencon.[4] Figure 4.9 shows the comparison of the experimental evaluation of N_γ^* for a rough surface foundation ($D_f = 0$) on a sand layer with theory. The angle of friction of the sand used for these tests was 43°. From Figure 4.9 the following conclusions can be drawn:

1. The value of N_γ^* for a given foundation increases with the decrease in H/B.
2. The magnitude of $H/B = D/B$ beyond which the presence of a rigid rough base has no influence on the N_γ^* value of a foundation is about 50%–75% more than that predicted by the theory.
3. For H/B between 0.5 to about 1.9, the experimental values of N_γ^* are higher than those predicted theoretically.
4. For $H/B <$ about 0.6, the experimental values of N_γ^* are lower than those predicted by theory. This may be due to two factors: (a) the crushing of sand grains at such high values of ultimate load, and (b) the curvilinear nature of the actual failure envelope of soil at high normal stress levels.

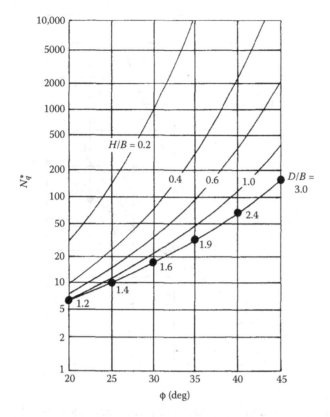

FIGURE 4.5 Mandel and Salencon's bearing capacity factor N_q^* [equation (4.2)].

Cerato and Lutenegger[9] reported laboratory model test results on large square and circular surface foundations. Based on these test results they observed that, at about $H/B \geq 3$,

$$N_\gamma^* \approx N_\gamma$$

Also, it was suggested that for surface foundations with $H/B < 3$,

$$q_u = 0.4\gamma B N_\gamma^* \quad \text{(square foundation)} \tag{4.6}$$

and

$$q_u = 0.3\gamma B N_\gamma^* \quad \text{(circular foundation)} \tag{4.7}$$

The variation of N_γ^* recommended by Cerato and Lutenegger[9] for use in equations (4.6) and (4.7) is given in Figure 4.10.

For saturated clay (that is, $\phi = 0$), equation (4.2) will simplify to the form

$$q_u = c_u N_c^* + q \tag{4.8}$$

Mandel and Salencon[10] performed calculations to evaluate N_c^* for *continuous foundations*. Similarly, Buisman[11] gave the following relationship for obtaining the

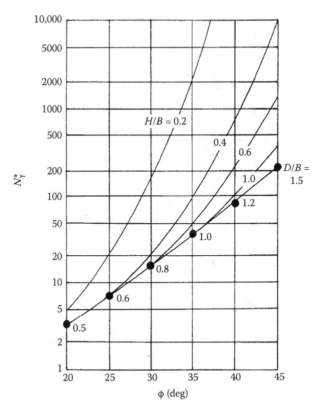

FIGURE 4.6 Mandel and Salencon's bearing capacity factor N_γ^* [equation (4.2)].

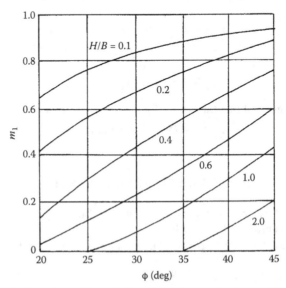

FIGURE 4.7 Variation of m_1 (Meyerhof's values) for use in the modified shape factor equation [equation (4.4)].

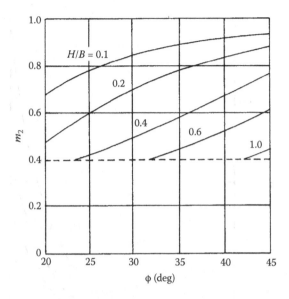

FIGURE 4.8 Variation of m_2 (Meyerhof's values) for use in equation (4.5).

ultimate bearing capacity of square foundations:

$$q_{u(\text{square})} = \left(\pi + 2 + \frac{B}{2H} - \frac{\sqrt{2}}{2} \right) c_u + q \quad \left(\text{for } \frac{B}{2H} - \frac{\sqrt{2}}{2} \geq 0 \right) \qquad (4.9)$$

where

c_u = undrained shear strength

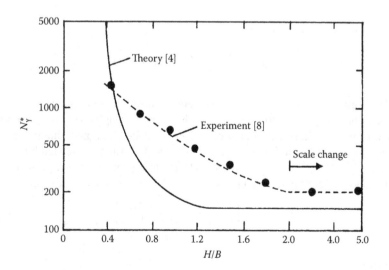

FIGURE 4.9 Comparison of theory with the experimental results of N_γ^* (*Note*: $\phi = 43°$, $c = 0$).

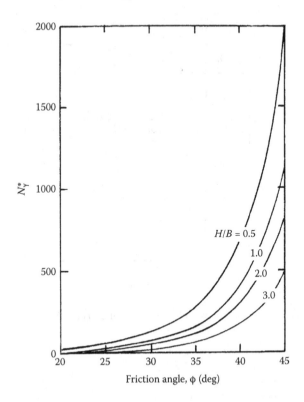

FIGURE 4.10 Cerato and Lutenegger's bearing capacity factor N_γ^* for use in equations (4.6) and (4.7). *Source:* Cerato, A. B., and A. J. Lutenegger. 2006. Bearing capacity of square and circular footings on a finite layer of granular soil underlain by a rigid base. *J. Geotech. Geoenv. Eng.*, ASCE, 132(11): 1496.

Equation (4.9) can be rewritten as

$$q_{u(\text{square})} = 5.14\left(1+\frac{0.5\frac{B}{H}-0.707}{5.14}\right)c_u + q = N_{c(\text{square})}^* c_u + q \qquad (4.10)$$

Table 4.1 gives the values of N_c^* for continuous and square foundations.

Equations (4.8) and (4.9) assume the existence of a rough rigid layer at a limited depth. However, if a soft saturated clay layer of limited thickness (undrained shear strength $= c_{u(1)}$) is located over another saturated clay with a somewhat larger shear strength $c_{u(2)}$ [Note: $c_{u(1)} < c_{u(2)}$; Figure 4.11], the following relationship suggested by Vesic[12] and DeBeer[13] may then be used to estimate the ultimate bearing capacity:

$$q_u = \left(1+0.2\frac{B}{L}\right)\left\{5.14+\left[1-\frac{c_{u(1)}}{c_{u(2)}}\right]\frac{\frac{B}{H}-\sqrt{2}}{2\left(\frac{B}{L}+1\right)}\right\}c_{u(1)} + q \qquad (4.11)$$

where
L = length of the foundation

TABLE 4.1

Values of N_c^* for Continuous and Square Foundations ($\phi = 0$ Condition)

$\dfrac{B}{H}$	N_c^*	
	Square[a]	Continuous[b]
2	5.43	5.24
3	5.93	5.71
4	6.44	6.22
5	6.94	6.68
6	7.43	7.20
8	8.43	8.17
10	9.43	9.05

[a] Buisman's analysis. *Source:* From Buisman, A. S. K. 1940. *Grond-mechanica.* Delft: Waltman.

[b] Mandel and Salencon's analysis. *Source:* From Mandel, J., and J. Salencon. 1969. Force portante d'un sol sur une assise rigide, in *Proc., VII Int. Conf. Soil Mech. Found Engg.*, Mexico City, 2, 157.

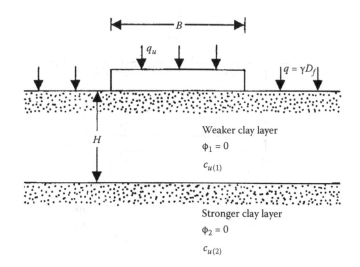

FIGURE 4.11 Foundation on a weaker clay underlain by a stronger clay layer (*Note:* $c_{u(1)} < c_{u(2)}$).

4.3 FOUNDATION ON LAYERED SATURATED ANISOTROPIC CLAY ($\phi = 0$)

Figure 4.12 shows a shallow continuous foundation supported by layered saturated anisotropic clay. The width of the foundation is B, and the interface between the clay layers is located at a depth H measured from the bottom of the foundation. It is assumed that the clays are anisotropic with respect to strength following the Casagrande-Carillo relationship,[14] or

$$c_{u(i)} = c_{u(h)} + [c_{u(v)} - c_{u(h)}]\sin^2 i \qquad (4.12)$$

where

$\quad c_{u(i)}$ = undrained shear strength at a given depth where the major principal stress is inclined at an angle i with the horizontal

$c_{u(v)}, c_{u(h)}$ = undrained shear strength for $i = 90°$ and $0°$, respectively

The ultimate bearing capacity of the continuous foundation can be given as

$$q_u = c_{u(v)-1}N_{c(L)} + q \qquad (4.13)$$

where

$\quad c_{u(v)-1}$ = undrained shear strength of the top soil layer when the major principal stress is vertical

$\quad q = \gamma_1 D_f$

$\quad D_f$ = depth of foundation

$\quad \gamma_1$ = unit weight of the top soil layer

$\quad N_{c(L)}$ = bearing capacity factor

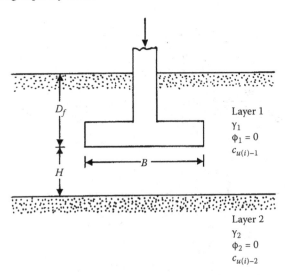

FIGURE 4.12 Shallow continuous foundation on layered anisotropic clay.

However, the bearing capacity factor $N_{c(L)}$ will be a function of H/B and $c_{u(v)-2}/c_{u(v)-1}$, or

$$N_{c(L)} = f\left[\frac{H}{B}, \frac{c_{u(v)-2}}{c_{u(v)-1}}\right] \qquad (4.14)$$

where
$\quad c_{u(v)-2} =$ undrained shear strength of the bottom clay layer when the major principal stress is vertical

Reddy and Srinivasan[15] developed a procedure to determine the variation of $N_{c(L)}$. In developing their theory, they assumed that the failure surface was cylindrical when the center of the trial failure surface was at O, as shown in Figure 4.13. They also assumed that the magnitudes of $c_{u(v)}$ for the top clay layer $[c_{u(v)-1}]$ and the bottom clay layer $[c_{u(v)-2}]$ remained constant with depth z as shown in Figure 4.13b.

For equilibrium of the foundation, considering forces per unit length and taking the moment about point O in Figure 4.13a,

$$2bq_u(r\sin\theta - b) = 2\int_{\theta_1}^{\theta} r^2[c_{u(i)-1}]d\alpha + 2\int_0^{\theta_1} r^2[c_{u(i)-2}]d\alpha \qquad (4.15)$$

where
$$b = \text{half-width of the foundation} = B/2$$
$$r = \text{radius of the trial failure circle}$$
$$c_{u(i)-1}, c_{u(i)-2} = \text{directional undrained shear strengths for layers 1 and 2, respectively}$$

As shown in Figure 4.13, let ψ be the angle between the failure plane and the direction of the major principal stress. Referring to equation (4.12):
Along arc AC,

$$c_{u(i)-1} = c_{u(h)-1} + [c_{u(v)-1} - c_{u(h)-1}]\sin^2(\alpha + \psi) \qquad (4.16)$$

Along arc CE,

$$c_{u(i)-2} = c_{u(h)-2} + [c_{u(v)-2} - c_{u(h)-2}]\sin^2(\alpha + \psi) \qquad (4.17)$$

Similarly, along arc DB,

$$c_{u(i)-1} = c_{u(h)-1} + [c_{u(v)-1} - c_{u(h)-1}]\sin^2(\alpha - \psi) \qquad (4.18)$$

and along arc ED,

$$c_{u(i)-2} = c_{u(h)-2} + [c_{u(v)-2} - c_{u(h)-2}]\sin^2(\alpha - \psi) \qquad (4.19)$$

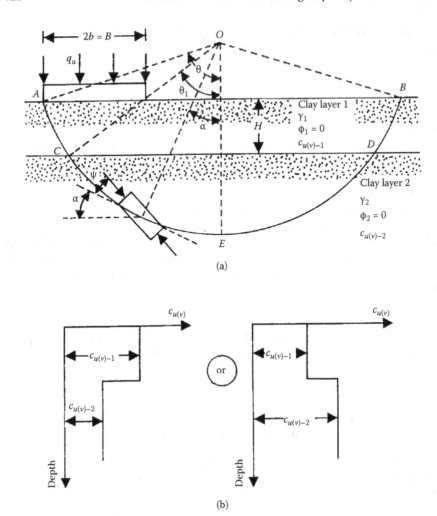

FIGURE 4.13 Assumptions in deriving $N_{c(L)}$ for a continuous foundation on anisotropic layered clay.

Note that $i = \alpha + \psi$ for the portion of the arc AE, and $i = \alpha - \psi$ for the portion BE. Let the anisotropy coefficient be defined as

$$K = \left[\frac{c_{u(v)-1}}{c_{u(h)-1}}\right] = \left[\frac{c_{u(v)-2}}{c_{u(h)-2}}\right] \qquad (4.20)$$

The magnitude of the anisotropy coefficient K is less than one for overconsolidated clays and $K > 1$ for normally consolidated clays. Also, let

$$n = \left[\frac{c_{u(v)-2}}{c_{u(v)-1}}\right] - 1 = \left[\frac{c_{u(h)-2}}{c_{u(h)-1}}\right] - 1 \qquad (4.21)$$

where

n = a factor representing the relative strength of two clay layers

Combining equations (4.15), (4.16), (4.17), (4.18), (4.19), and (4.20),

$$2bq_u(r\sin\theta - b) = \int_{\theta_1}^{\theta} r^2 \left\{ c_{u(h)-1} + [c_{u(v)-1} - c_{u(h)-1}]\sin^2(\alpha + \psi) \right\} d\alpha$$

$$+ \int_{\theta_1}^{\theta} r^2 \left\{ c_{u(h)-1} + [c_{u(v)-1} - c_{u(h)-1}]\sin^2(\alpha - \psi) \right\} d\alpha \qquad (4.22)$$

$$+ \int_{0}^{\theta_1} r^2(n+1) \left\{ c_{u(h)-1} + [c_{u(v)-1} - c_{u(h)-1}]\sin^2(\alpha + \psi) \right\} d\alpha$$

$$+ \int_{0}^{\theta_1} r^2(n+1) \left\{ c_{u(h)-1} + [c_{u(v)-1} - c_{u(h)-1}]\sin^2(\alpha - \psi) \right\} d\alpha$$

Or, combining equations (4.20) and (4.22),

$$\frac{q_u}{c_{u(v)-1}} = \frac{\frac{r^2}{b^2}}{2K\left[\left(\frac{r}{b}\right)\sin\theta - 1\right]} \left\{ \begin{array}{l} 2\theta + 2n\theta_1 + (K-1)\theta + n(K-1)\theta_1 \\[2mm] -\dfrac{K-1}{2}\left[\dfrac{\sin 2(\theta + \psi)}{2} + \dfrac{\sin 2(\theta - \psi)}{2}\right] \\[2mm] -\dfrac{n(K-1)}{2}\left[\dfrac{\sin 2(\theta_1 + \psi)}{2} + \dfrac{\sin 2(\theta_1 - \psi)}{2}\right] \end{array} \right\} \qquad (4.23)$$

where

$$\theta_1 = \cos^{-1}\left(\cos\theta + \frac{H}{r}\right)$$

From equation (4.13) note that, with $q = 0$ (surface foundation),

$$N_{c(L)} = \frac{q_u}{c_{u(v)-1}} \qquad (4.24)$$

In order to obtain the minimum value of $N_{c(L)} = q_u/c_{u(v)-1}$, the theorem of maxima and minima needs to be used, or

$$\frac{\partial N_{c(L)}}{\partial \theta} = 0 \qquad (4.25)$$

and

$$\frac{\partial N_{c(L)}}{\partial r} = 0 \qquad (4.26)$$

Equations (4.23), (4.25), and (4.26) will yield two relationships in terms of the variables θ and r/b. So, for given values of H/b, K, n, and ψ, the above relationships may be solved to obtain values of θ and r/b. These can then be used in equation (4.23) to obtain the desired value of $N_{c(L)}$ (for given values of H/b, K, n, and ψ). Lo[16] showed that the angle ψ between the failure plane and the major principal stress for anisotropic soils can be taken to be approximately equal to 35°. The variations of the bearing capacity factor $N_{c(L)}$ obtained in this manner for $K = 0.8$, 1 (isotropic case), 1.2, 1.4, 1.6, and 1.8, are shown in Figure 4.14.

If a shallow rectangular foundation $B \times L$ in plan is located at a depth D_f, the general ultimate bearing capacity equation [see equation (2.82)] will be of the form ($\phi = 0$ condition)

$$q_u = c_{u(v)-1} N_{c(L)} \lambda_{cs} \lambda_{cd} + q \lambda_{qs} \lambda_{qd} \tag{4.27}$$

where

λ_{cs}, λ_{qs} = shape factors
λ_{cd}, λ_{qd} = depth factors

The proper shape and depth factors can be selected from Table 2.6.

EXAMPLE 4.1

Refer to Figure 4.12. For the foundation, given: $D_f = 0.8$ m; $B = 1$ m; $L = 1.6$ m; $H = 0.5$ m; $\gamma_1 = 17.8$ kN/m³; $\gamma_2 = 17.0$ kN/m³; $c_{u(v)-1} = 45$ kN/m²; $c_{u(v)-2} = 30$ kN/m²; anisotropy coefficient $K = 1.4$. Estimate the allowable load-bearing capacity of the foundation with a factor of safety $FS = 4$. Use Meyerhof's shape and depth factors (Table 2.6).

Solution

From equation (4.27),

$$q_u = c_{u(v)-1} N_{c(L)} \lambda_{cs} \lambda_{cd} + q \lambda_{qs} \lambda_{qd}$$

$$\frac{H}{b} = \frac{H}{\left(\dfrac{B}{2}\right)} = \frac{0.5}{0.5} = 1; K = 1.4$$

$$\frac{c_{u(v)-2}}{c_{u(v)-1}} = \frac{30}{45} = 0.67$$

$$n = 1 - 0.67 = 0.33$$

So, from Figure 4.14(d), the value of $N_{c(L)} = 4.75$.
Using Meyerhof's shape and depth factors given in Table 2.6,

$$\lambda_{cs} = 1 + 0.2 \left(\frac{B}{L}\right) = 1 + (0.2)\left(\frac{1}{1.6}\right) = 1.125$$

$$\lambda_{qs} = 1$$

$$\lambda_{cd} = 1 + 0.2 \left(\frac{D_f}{B}\right) = 1 + (0.2)\left(\frac{0.8}{1.0}\right) = 1.16$$

$$\lambda_{qd} = 1$$

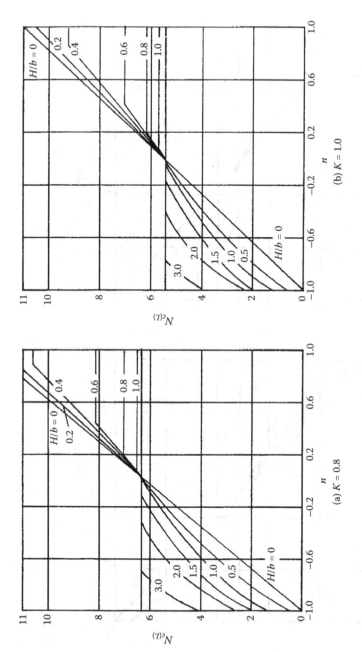

FIGURE 4.14 Bearing capacity factor $N_{c(L)}$. *Source:* From Reddy, A. S., and R. J. Srinivasan. 1967. Bearing capacity of footings on layered clays. *J. Soil Mech. Found. Div.*, ASCE, 93(SM2): 83.

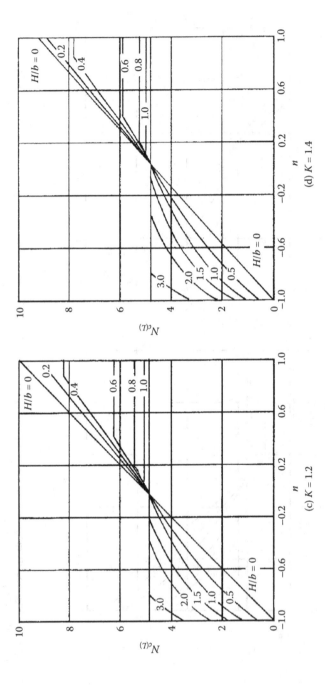

FIGURE 4.14 (Continued)

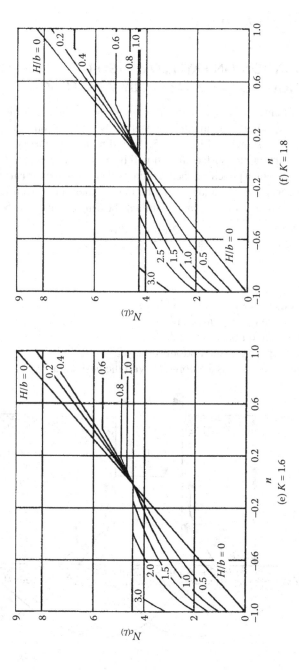

FIGURE 4.14 (Continued)

So,

$$q_u = (45)(4.75)(1.125)(1.16) + (17.8)(0.8)(1.0)(1.0) = 278.9 + 14.24 = 293.14 \text{ kN/m}^2$$

$$q_{all} = \frac{q_u}{FS} = \frac{293.14}{4} = \textbf{73.29 kN/m}^2$$

4.4   FOUNDATION ON LAYERED $c - \phi$ SOIL— STRONGER SOIL UNDERLAIN BY WEAKER SOIL

Meyerhof and Hanna[17] developed a theory to estimate the ultimate bearing capacity of a shallow rough continuous foundation supported by a strong soil layer underlain by a weaker soil layer as shown in Figure 4.15. According to their theory, at ultimate load per unit area q_u, the failure surface in soil will be as shown in Figure 4.15. If the ratio H/B is relatively small, a punching shear failure will occur in the top (stronger) soil layer followed by a general shear failure in the bottom (weaker) layer. Considering the unit length of the continuous foundation, the ultimate bearing capacity can be given as

$$q_u = q_b + \frac{2(C_a + P_p \sin \delta)}{B} - \gamma_1 H \tag{4.28}$$

where
     B = width of the foundation
     γ_1 = unit weight of the stronger soil layer
     C_a = adhesive force along aa' and bb'
     P_p = passive force on faces aa' and bb'
     q_b = bearing capacity of the bottom soil layer
     δ = inclination of the passive force P_p with the horizontal

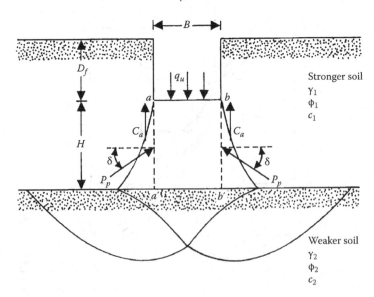

FIGURE 4.15 Rough continuous foundation on layered soil—stronger over weaker.

Note that, in equation (4.28),

$$C_a = c_a H \tag{4.29}$$

where

c_a = unit adhesion

$$P_p = \frac{1}{2}\gamma_1 H^2 \left(\frac{K_{pH}}{\cos\delta}\right) + (\gamma_1 D_f)(H)\left(\frac{K_{pH}}{\cos\delta}\right) = \frac{1}{2}\gamma_1 H^2\left(1+\frac{2D_f}{H}\right)\left(\frac{K_{pH}}{\cos\delta}\right) \tag{4.30}$$

where

K_{pH} = horizontal component of the passive earth pressure coefficient

Also,

$$q_b = c_2 N_{c(2)} + \gamma_1(D_f + H)N_{q(2)} + \tfrac{1}{2}\gamma_2 BN_{\gamma(2)} \tag{4.31}$$

where

c_2 = cohesion of the bottom (weaker) layer of soil
γ_2 = unit weight of bottom soil layer
$N_{c(2)}, N_{q(2)}, N_{\gamma(2)}$ = bearing capacity factors for the bottom soil layer (that is, with respect to the soil friction angle of the bottom soil layer ϕ_2)

Combining equations (4.28), (4.29), and (4.30),

$$q_u = q_b + \frac{2c_a H}{B} + 2\left[\frac{1}{2}\gamma_1 H^2\left(1+\frac{2D_f}{H}\right)\right]\left(\frac{K_{pH}}{\cos\delta}\right)\left(\frac{\sin\delta}{B}\right) - \gamma_1 H$$

$$= q_b + \frac{2c_a H}{B} + \gamma_1 H^2\left(1+\frac{2D_f}{H}\right)\frac{K_{pH}\tan\delta}{B} - \gamma_1 H \tag{4.32}$$

Let

$$K_{pH}\tan\delta = K_s \tan\phi_1 \tag{4.33}$$

where

K_s = punching shear coefficient

So,

$$q_u = q_b + \frac{2c_a H}{B} + \gamma_1 H^2\left(1+\frac{2D_f}{H}\right)\frac{K_s \tan\phi_1}{B} - \gamma_1 H \tag{4.34}$$

The punching shear coefficient can be determined using the passive earth pressure coefficient charts proposed by Caquot and Kerisel.[18] Figure 4.16 gives the variation of K_s with q_2/q_1 and ϕ_1. Note that q_1 and q_2 are the ultimate bearing capacities of a continuous surface foundation of width B under vertical load on homogeneous beds of upper and lower soils, respectively, or

$$q_1 = c_1 N_{c(1)} + \tfrac{1}{2}\gamma_1 BN_{\gamma(1)} \tag{4.35}$$

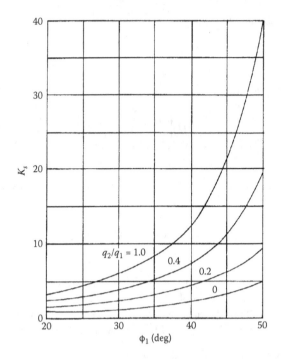

FIGURE 4.16 Meyerhof and Hanna's theory—variation of K_s with ϕ_1 and q_2/q_1.

where

$N_{c(1)}, N_{\gamma(1)}$ = bearing capacity factors corresponding to soil friction angle ϕ_1

$$q_2 = c_2 N_{c(2)} + \tfrac{1}{2}\gamma_2 BN_{\gamma(2)} \qquad (4.36)$$

If the height H is large compared to the width B (Figure 4.15), then the failure surface will be completely located in the upper stronger soil layer, as shown in Figure 4.17. In such a case, the upper limit for q_u will be of the following form:

$$q_u = q_t = c_1 N_{c(1)} + qN_{q(1)} + \tfrac{1}{2}\gamma_1 BN_{\gamma(1)} \qquad (4.37)$$

Hence, combining equations (4.34) and (4.37),

$$q_u = q_b + \frac{2c_a H}{B} + \gamma_1 H^2\left(1 + \frac{2D_f}{H}\right)\frac{K_s \tan \phi_1}{B} - \gamma_1 H \le q_t \qquad (4.38)$$

For rectangular foundations, the preceding equation can be modified as

$$q_u = q_b + \left(1 + \frac{B}{L}\right)\left(\frac{2c_a H}{B}\right)\lambda_a + \left(1 + \frac{B}{L}\right)\gamma_1 H^2\left(1 + \frac{2D_f}{H}\right)\left(\frac{K_s \tan \phi_1}{B}\right)\lambda_s - \gamma_1 H \le q_t \qquad (4.39)$$

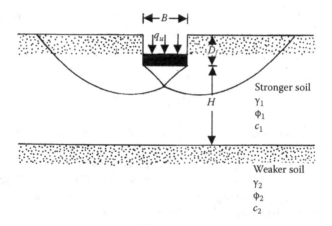

FIGURE 4.17 Continuous rough foundation on layered soil—H/B is relatively small.

where

$$\lambda_a, \lambda_s = \text{shape factors}$$

$$q_b = c_2 N_{c(2)} \lambda_{cs(2)} + \gamma_1 (D_f + H) N_{q(2)} \lambda_{qs(2)} + \tfrac{1}{2} \gamma_2 B N_{\gamma(2)} \lambda_{\gamma s(2)} \quad (4.40)$$

$$q_t = c_1 N_{c(1)} \lambda_{cs(1)} + \gamma_1 D_f N_{q(1)} \lambda_{qs(1)} + \tfrac{1}{2} \gamma_1 B N_{\gamma(1)} \lambda_{\gamma s(1)} \quad (4.41)$$

$\lambda_{cs(1)}, \lambda_{qs(1)}, \lambda_{\gamma s(1)} = $ shape factors for the top soil layer (friction angle $= \phi_1$; see Table 2.6)

$\lambda_{cs(2)}, \lambda_{qs(2)}, \lambda_{\gamma s(2)} = $ shape factors for the bottom soil layer (friction angle $= \phi_2$; see Table 2.6)

Based on the general equations [equations (4.39), (4.40), and (4.41)], some special cases may be developed. They are as follows:

CASE I: STRONGER SAND LAYER OVER WEAKER SATURATED CLAY ($\phi_2 = 0$)

For this case, $c_1 = 0$; hence, $c_a = 0$. Also for $\phi_2 = 0$, $N_{c(2)} = 5.14$, $N_{\gamma(2)} = 0$, $N_{q(2)} = 1$, $\lambda_{cs} = 1 + 0.2(B/L)$, $\lambda_{qs} = 1$ (shape factors are Meyerhof's values as given in Table 2.6). So,

$$q_u = 5.14 c_2 \left[1 + 0.2 \left(\frac{B}{L} \right) \right] + \left(1 + \frac{B}{L} \right) \gamma_1 H^2 \left(1 + \frac{2D_f}{H} \right) \left(\frac{K_s \tan \phi_1}{B} \right) \lambda_s + \gamma_1 D_f \le q_t \quad (4.42)$$

where

$$q_t = \gamma_1 D_f N_{q(1)} \left[1 + 0.1 \left(\frac{B}{L} \right) \tan^2 \left(45 + \frac{\phi_1}{2} \right) \right] + \frac{1}{2} \gamma_1 B N_{\gamma(1)} \left[1 + 0.1 \left(\frac{B}{L} \right) \tan^2 \left(45 + \frac{\phi_1}{2} \right) \right] \quad (4.43)$$

In equation (4.43) the relationships for the shape factors λ_{qs} and $\lambda_{\gamma s}$ are those given by Meyerhof[19] as shown in Table 2.6. Note that K_s is a function of q_2/q_1 [equations (4.35) and (4.36)]. For this case,

$$\frac{q_2}{q_1} = \frac{c_2 N_{c(2)}}{\frac{1}{2}\gamma_1 BN_{\gamma(1)}} = \frac{5.14c_2}{0.5\gamma_1 BN_{\gamma(1)}} \tag{4.44}$$

Once q_2/q_1 is known, the magnitude of K_s can be obtained from Figure 4.16, which, in turn, can be used in equation (4.42) to determine the ultimate bearing capacity of the foundation q_u. The value of the shape factor λ_s for a strip foundation can be taken as one. As per the experimental work of Hanna and Meyerhof,[20] the magnitude of λ_s appears to vary between 1.1 and 1.27 for square or circular foundations. For conservative designs, it may be taken as one.

Based on this concept, Hanna and Meyerhof[20] developed some alternative design charts to determine the punching shear coefficient K_s, and these charts are shown in Figures 4.18 and 4.19. In order to use these charts, the ensuing steps need to be followed.

1. Determine q_2/q_1.
2. With known values of ϕ_1 and q_2/q_1, determine the magnitude of δ/ϕ_1 from Figure 4.18.
3. With known values of ϕ_1, δ/ϕ_1, and c_2, determine K_s from Figure 4.19.

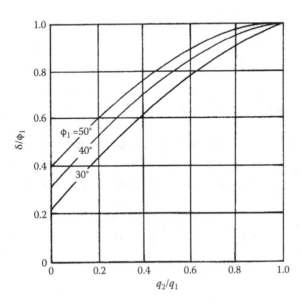

FIGURE 4.18 Hanna and Meyerhof's analysis—variation of δ/ϕ_1 with ϕ_1 and q_2/q_1—stronger sand over weaker clay.

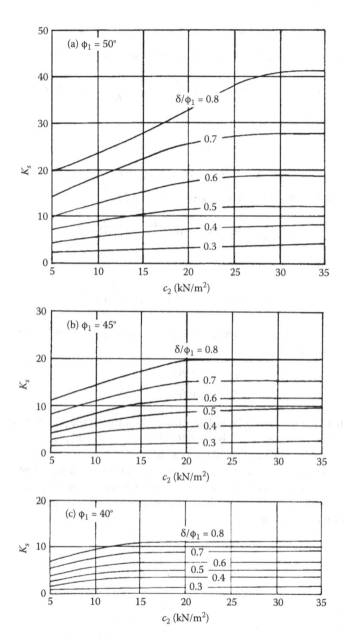

FIGURE 4.19 Hanna and Meyerhof's analysis for coefficient of punching shear—stronger sand over weaker clay.

CASE II: STRONGER SAND LAYER OVER WEAKER SAND LAYER

For this case, $c_1 = 0$ and $c_a = 0$. Hence, referring to equation (4.39),

$$q_u = q_b + \left(1+\frac{B}{L}\right)\gamma_1 H^2\left(1+\frac{2D_f}{H}\right)\left(\frac{K_s\tan\phi_1}{B}\right)\lambda_s - \gamma_1 H \le q_t \tag{4.45}$$

where

$$q_b = \gamma_1(D_f+H)N_{q(2)}\lambda_{qs(2)} + \tfrac{1}{2}\gamma_2 BN_{\gamma(2)}\lambda_{\gamma s(2)} \tag{4.46}$$

$$q_t = \gamma_1 D_f N_{q(1)}\lambda_{qs(1)} + \tfrac{1}{2}\gamma_1 BN_{\gamma(1)}\lambda_{\gamma s(1)} \tag{4.47}$$

Using Meyerhof's shape factors given in Table 2.6,

$$\lambda_{qs(1)} = \lambda_{\gamma s(1)} = 1+0.1\left(\frac{B}{L}\right)\tan^2\left(45+\frac{\phi_1}{2}\right) \tag{4.48}$$

and

$$\lambda_{qs(2)} = \lambda_{\gamma s(2)} = 1+0.1\left(\frac{B}{L}\right)\tan^2\left(45+\frac{\phi_2}{2}\right) \tag{4.49}$$

For conservative designs, for all B/L ratios, the magnitude of λ_s can be taken as one. For this case

$$\frac{q_2}{q_1} = \frac{0.5\gamma_2 BN_{\gamma(2)}}{0.5\gamma_1 BN_{\gamma(1)}} = \frac{\gamma_2 N_{\gamma(2)}}{\gamma_1 N_{\gamma(1)}} \tag{4.50}$$

Once the magnitude of q_2/q_1 is determined, the value of the punching shear coefficient K_s can be obtained from Figure 4.16. Hanna[21] suggested that the friction angles obtained from direct shear tests should be used.

Hanna[21] also provided an improved design chart for estimating the punching shear coefficient K_s in equation (4.45). In this development he assumed that the variation of δ for the assumed failure surface in the top stronger sand layer will be of the nature shown in Figure 4.20, or

$$\delta_{z'} = \eta\phi_2 + az'^2 \tag{4.51}$$

where

$$\eta = \frac{q_2}{q_1} \tag{4.52}$$

$$a = \frac{\phi_1 - \left(\frac{q_2}{q_1}\right)\phi_2}{H^2} \tag{4.53}$$

So,

$$\delta_{z'} = \left(\frac{q_2}{q_1}\right)\phi_2 + \left[\frac{\phi_1 - \left(\frac{q_2}{q_1}\right)\phi_2}{H^2}\right]z'^2 \tag{4.54}$$

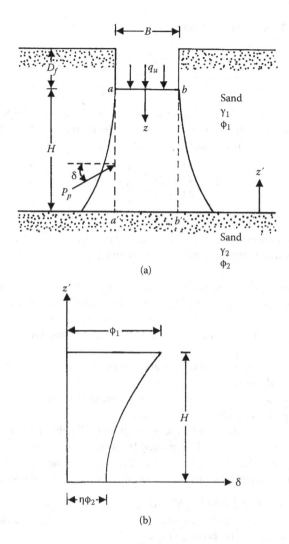

(a)

(b)

FIGURE 4.20 Hanna's assumption for variation of δ with depth for determination of K_s.

The preceding relationship means that at $z' = 0$ (that is, at the interface of the two soil layers),

$$\delta = \left(\frac{q_2}{q_1} \right) \phi_2 \tag{4.55}$$

and at the level of the foundation, that is $z' = H$,

$$\delta = \phi_1 \tag{4.56}$$

Equation (4.51) can also be rewritten as

$$\delta_z = \left(\frac{q_2}{q_1}\right)\phi_2 + \left[\frac{\phi_1 - \left(\frac{q_2}{q_1}\right)\phi_2}{H^2}\right](H-z)^2 \qquad (4.57)$$

where δ_z is the angle of inclination of the passive pressure with respect to the horizontal at a depth z measured from the bottom of the foundation. So, the passive force per unit length of the vertical surface aa' (or bb') is

$$P_p = \int_0^H \left[\frac{\gamma_1 K_{pH(z)}}{\cos\delta_z}\right](z+D_f)dz \qquad (4.58)$$

where
   $K_{pH(z)}$ = horizontal component of the passive earth pressure coefficient at a depth z measured from the bottom of the foundation

The magnitude of P_p expressed by equation (4.58), in combination with the expression δ_z given in equation (4.57), can be determined. In order to determine the magnitude of the punching shear coefficient K_s given in equation (4.33), we need to know an *average* value of δ. In order to achieve that, the following steps are taken:

1. Assume an average value of δ and obtain K_{pH} as given in the tables by Caquot and Kerisel.[18]
2. Using the average values of δ and K_{pH} obtained from step 1, calculate P_p from equation (4.30).
3. Repeat steps 1 and 2 until the magnitude of P_p obtained from equation (4.30) is the same as that calculated from equation (4.58).
4. The average value of δ [for which P_p calculated from equations (4.30) and (4.58) is the same] is the value that needs to be used in equation (4.33) to calculate K_s.

Figure 4.21 gives the relationship for δ/ϕ_1 versus ϕ_2 for various values of ϕ_1 obtained by the above procedure. Using Figure 4.21, Hanna[21] gave a design chart for K_s, and this design chart is shown in Figure 4.22.

CASE III: STRONGER CLAY LAYER ($\phi_1 = 0$) OVER WEAKER CLAY ($\phi_2 = 0$)

For this case, $N_{q(1)}$ and $N_{q(2)}$ are both equal to one and $N_{\gamma(1)} = N_{\gamma(2)} = 0$. Also, $N_{c(1)} = N_{c(2)} = 5.14$. So, from equation (4.39),

$$q_u = \left[1 + 0.2\left(\frac{B}{L}\right)\right]c_2 N_{c(2)} + \left(1 + \frac{B}{L}\right)\left(\frac{2c_a H}{B}\right)\lambda_a + \gamma_1 D_f \le q_t \qquad (4.59)$$

where

$$q_t = \left[1 + 0.2\left(\frac{B}{L}\right)\right]c_1 N_{c(1)} + \gamma_1 D_f \qquad (4.60)$$

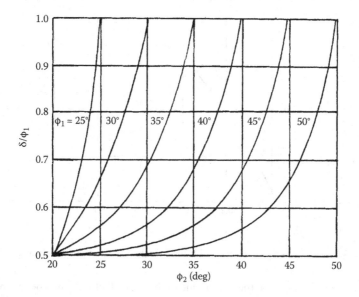

FIGURE 4.21 Hanna's analysis—variation of δ/ϕ_1.

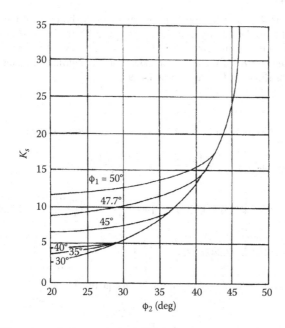

FIGURE 4.22 Hanna's analysis—variation of K_s for stronger sand over weaker sand.

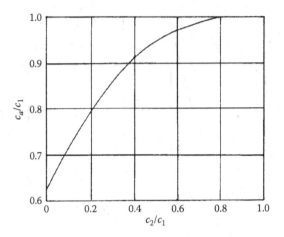

FIGURE 4.23 Analysis of Meyerhof and Hanna for the variation of c_a/c_1 with c_2/c_1.

For conservative design the magnitude of the shape factor λ_a may be taken as one. The magnitude of the adhesion c_a is a function of q_2/q_1. For this condition,

$$\frac{q_2}{q_1} = \frac{c_2 N_{c(2)}}{c_1 N_{c(1)}} = \frac{5.14 c_2}{5.14 c_1} = \frac{c_2}{c_1} \tag{4.61}$$

Figure 4.23 shows the theoretical variation of c_a with q_2/q_1.[17]

EXAMPLE 4.2

Refer to Figure 4.15. Let the top layer be sand and the bottom layer saturated clay. Given: $H = 1.5$ m. For the top layer (sand): $\gamma_1 = 17.5$ kN/m³; $\phi_1 = 40°$; $c_1 = 0$; for the bottom layer (saturated clay): $\gamma_2 = 16.5$ kN/m³; $\phi_2 = 0$; $c_2 = 30$ kN/m²; and for the foundation (continuous): $B = 2$ m; $D_f = 1.2$ m. Determine the ultimate bearing capacity q_u. Use the results shown in Figures 4.18 and 4.19.

Solution

For the continuous foundation $\frac{B}{L} = 0$ and $\lambda_s = 1$, in equation 4.42 we obtain

$$q_u = 5.14 c_2 + \gamma_1 H^2 \left(1 + \frac{2D_f}{H}\right)\left(\frac{K_s \tan\phi_1}{B}\right) + \gamma_1 D_f$$

$$= (5.14)(30) + (17.5)(H^2)\left[1 + \frac{(2)(1.2)}{H}\right]\frac{K_s \tan 40}{2} + (17.5)(1.2)$$

$$= 175.2 + 7.342 H^2 K_s \left(1 + \frac{2.4}{H}\right) \tag{a}$$

To determine K_s, we need to obtain q_2/q_1. From equation (4.44),

$$\frac{q_2}{q_1} = \frac{5.14 c_2}{0.5 \gamma B N_{\gamma(1)}}$$

From Table 2.3 for $\phi_1 = 40°$, Meyerhof's value of $N_{\gamma(1)}$ is equal to 93.7. So,

$$\frac{q_2}{q_1} = \frac{(5.14)(30)}{(0.5)(17.5)(2)(93.7)} = 0.094$$

Referring to Figure 4.18 for $q_2/q_1 = 0.094$ and $\phi_1 = 40°$, the value of $\delta/\phi_1 = 0.42$. With $\delta/\phi_1 = 0.42$ and $c_2 = 30$ kN/m², Figure 4.19c gives the value of $K_s = 3.89$. Substituting this value into equation (a) gives

$$q_u = 175.2 + 28.56H^2\left(1 + \frac{2.4}{H}\right) \le q_t \tag{b}$$

From equation (4.43),

$$q_t = \gamma_1 D_f N_{q(1)}\left[1 + 0.1\left(\frac{B}{L}\right)\tan^2\left(45 + \frac{\phi_1}{2}\right)\right] + \frac{1}{2}\gamma_1 BN_{\gamma(1)}\left[1 + 0.1\left(\frac{B}{L}\right)\tan^2\left(45 + \frac{\phi_1}{2}\right)\right] \tag{c}$$

For the continuous foundation $B/L = 0$. So,

$$q_t = \gamma_1 D_f N_{q(1)} + \frac{1}{2}\gamma_1 BN_{\gamma(1)}$$

For $\phi_1 = 40°$, use Meyerhof's values of $N_{\gamma(1)} = 93.7$ and $N_{q(1)} = 62.4$ (Table 2.3). Hence,

$$q_t = (17.5)(1.2)(62.4) + \frac{1}{2}(17.5)(2)(93.7) = 1348.2 + 1639.75 = 2987.95 \text{ kN/m}^2$$

If $H = 1.5$ m is substituted into equation (b),

$$q_u = 175.2 + (28.56)(1.5)^2\left(1 + \frac{2.4}{1.5}\right) = 342.3 \text{ kN/m}^2$$

Since $q_u = 342.3 < q_t$, the ultimate bearing capacity is **342.3 kN/m²**

EXAMPLE 4.3

Refer to Figure 4.15, which shows a square foundation on layered sand. Given: $H = 1.0$ m. Also given for the top sand layer: $\gamma_1 = 18$ kN/m³; $\phi_1 = 40°$; for the bottom sand layer: $\gamma_2 = 16.5$ kN/m³; $\phi_2 = 32°$; and for the foundation: $B \times B = 1.5$ m × 1.5 m; $D_f = 1.5$ m. Estimate the ultimate bearing capacity of the foundation. Use Figure 4.22.

Solution

From equation (4.45),

$$q_u = q_b + \left(1 + \frac{B}{L}\right)\gamma_1 H^2\left(1 + \frac{2D_f}{H}\right)\left(\frac{K_s \tan\phi_1}{B}\right)\lambda_s - \gamma_1 H \le q_t$$

$$\lambda_s \approx 1$$

Given: $\phi_1 = 40°$; $\phi_2 = 32°$. From Figure 4.22, $K_s \approx 5.75$. From equation (4.46),

$$q_b = \gamma_1(D_f + H)N_{q(2)}\lambda_{qs(2)} + \frac{1}{2}\gamma_2 BN_{\gamma(2)}\lambda_{\gamma s(2)}$$

For $\phi_2 = 32°$, Meyerhof's bearing capacity factors are $N_{\gamma(2)} = 22.02$ and $N_{q(2)} = 23.18$ (Table 2.3). Also from Table 2.6, Meyerhof's shape factors

$$\lambda_{qs(2)} = \lambda_{\gamma s(2)} = 1 + 0.1\left(\frac{B}{L}\right)\tan^2\left(45 + \frac{\phi_2}{2}\right) = 1 + (0.1)\left(\frac{1.5}{1.5}\right)\tan^2\left(45 + \frac{32}{2}\right) = 1.325$$

$$q_b = (18)(1.5+1)(23.18)(1.325) + \frac{1}{2}(16.7)(1.5)(22.02)(1.325) = 1382.1 + 365.4 = 1747.5 \text{ kN/m}^2$$

Hence, from equation (4.45),

$$q_u = 1747.5 + \left(1 + \frac{1.5}{1.5}\right)(18)(1)^2\left(1 + \frac{2 \times 1.5}{1}\right)\left(\frac{5.75\tan 40}{1.5}\right) - (18)(1) = 1747.5 + 463.2 - 18$$

$$= 2192.7 \text{ kN/m}^2$$

CHECK

From equation (4.47),

$$q_t = \gamma_1(D_f + H)N_{q(1)}\lambda_{qs(1)} + \tfrac{1}{2}\gamma_1 BN_{\gamma(1)}\lambda_{\gamma s(2)}$$

For $\phi_1 = 40°$, Meyerhof's bearing capacity factors are $N_{q(1)} = 62.4$ and $N_{\gamma(1)} = 93.69$ (Table 2.3).

$$\lambda_{qs(1)} = \lambda_{\gamma s(1)} = 1 + 0.1\left(\frac{B}{L}\right)\tan^2\left(45 + \frac{\phi_1}{2}\right) = 1 + (0.1)\left(\frac{1.5}{1.5}\right)\tan^2\left(45 + \frac{40}{2}\right) \approx 1.46$$

$$q_t = (18)(1.5+1)(64.2)(1.46) + \tfrac{1}{2}(18)(1.5)(93.69)(1.46) = 4217.9 + 1846.6 = 6064.5 \text{ kN/m}^2$$

So, $q_u = \textbf{2192.7 kN/m}^2$.

EXAMPLE 4.4

Figure 4.24 shows a shallow foundation. Given: $H = 1$ m; undrained shear strength c_1 (for $\phi_1 = 0$ condition) $= 80$ kN/m^2; undrained shear strength c_2 (for $\phi_2 = 0$ condition) $= 32$ kN/m^2; $\gamma_1 = 18$ kN/m^3; $D_f = 1$ m; $B = 1.5$ m; $L = 3$ m. Estimate the ultimate bearing capacity of the foundation.

Solution

From equation (4.61),

$$\frac{q_2}{q_1} = \frac{c_2}{c_1} = \frac{32}{80} = 0.4$$

From Figure 4.23 for $q_2/q_1 = 0.4$, $c_a/c_1 = 0.9$. So $c_a = (0.9)(80) = 72$ kN/m^2. From equation (4.60),

$$q_t = \left[1 + 0.2\left(\frac{B}{L}\right)\right]c_1 N_{c(1)} + \gamma_1 D_f = \left[1 + 0.2\left(\frac{1.5}{3}\right)\right](80)(5.14) + (18)(1) = 470.32 \text{ kN/m}^2.$$

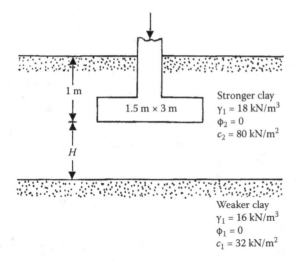

FIGURE 4.24 Shallow foundation on layered clay.

With $\lambda_s = 1$, equation (4.59) yields

$$q_u = \left[1 + 0.2\left(\frac{B}{L}\right)\right]c_2 N_{c(2)} + \left(1 + \frac{B}{L}\right)\left(\frac{2c_a H}{B}\right)\lambda_a + \gamma_1 D_f \le q_t$$

$$= (1 + 0.1)(32)(5.14) + (1.5)\left[(2)(72)\left(\frac{H}{B}\right)\right] + (18)(1)$$

$$= 198.93 + 216\left(\frac{H}{B}\right) = 198.93 + 216\left(\frac{1}{1.5}\right) \approx \mathbf{343\ kN/m^2}$$

4.5 FOUNDATION ON LAYERED SOIL—WEAKER SOIL UNDERLAIN BY STRONGER SOIL

In general, when a foundation is supported by a weaker soil layer underlain by stronger soil at a shallow depth as shown in the left-hand side of Figure 4.25, the failure surface at ultimate load will pass through both soil layers. However, when the magnitude of H is relatively large compared to the width of the foundation B, the failure surface at ultimate load will be fully located in the weaker soil layer (see the right-hand side of Figure 4.25). The procedure to estimate the ultimate bearing capacity of such foundations on layered sand and layered saturated clay follows.

4.5.1 FOUNDATIONS ON WEAKER SAND LAYER UNDERLAIN BY STRONGER SAND ($c_1 = 0$, $c_2 = 0$)

Based on several laboratory model tests, Hanna[22] proposed the following relationship for estimating the ultimate bearing capacity q_u for a foundation resting on a

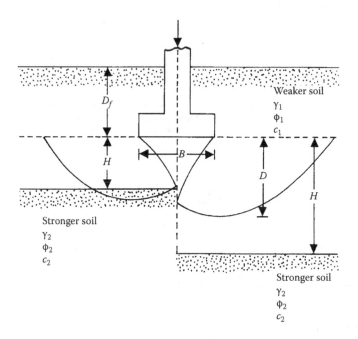

FIGURE 4.25 Foundation on weaker soil layer underlain by stronger sand layer.

weak sand layer underlain by a strong sand layer:

$$q_u = \tfrac{1}{2}\gamma_1 \lambda^*_{\gamma s} N_{\gamma(m)} + \gamma_1 \lambda^*_{qs} N_{q(m)} \leq \tfrac{1}{2}\gamma_2 \lambda_{\gamma s(2)} N_{\gamma(2)} + \gamma_2 D_f \lambda_{qs(2)} N_{q(2)} \qquad (4.62)$$

where

$\quad N_{\gamma(2)}, N_{q(2)}$ = Meyerhof's bearing capacity factors with reference to soil friction
$\qquad$ angle ϕ_2 (Table 2.3)

$\quad \lambda_{\gamma s(2)}, \lambda_{qs(2)}$ = Meyerhof's shape factors (Table 2.6) with reference to soil friction

$\qquad$ angle $\phi_2 = 1 + 0.1\left(\dfrac{B}{L}\right)\tan^2\left(45 + \dfrac{\phi_2}{2}\right)$

$\quad N_{\gamma(m)}, N_{q(m)}$ = modified bearing capacity factors
$\quad \lambda^*_{\gamma s}, \lambda^*_{\gamma s}$ = modified shape factors

The modified bearing capacity factors can be obtained as follows:

$$N_{\gamma(m)} = N_{\gamma(2)} - \left[\frac{H}{D_{(\gamma)}}\right][N_{\gamma(2)} - N_{\gamma(1)}] \qquad (4.63)$$

$$N_{q(m)} = N_{q(2)} - \left[\frac{H}{D_{(q)}}\right][N_{q(2)} - N_{q(1)}] \qquad (4.64)$$

where

$N_{\gamma(1)}, N_{q(1)}$ = Meyerhof's bearing capacity factors with reference to soil friction angle ϕ_1 (Table 2.3)

The variations of $D_{(\gamma)}$ and $D_{(q)}$ with ϕ_1 are shown in Figures 4.2 and 4.3. The relationships for the modified shape factors are the same as those given in equations (4.4) and (4.5). The term m_1 [equation (4.4)] can be determined from Figure 4.7 by substituting $D_{(q)}$ for H and ϕ_1 for ϕ. Similarly, the term m_2 [equation (4.5)] can be determined from Figure 4.8 by substituting $D_{(\gamma)}$ for H and ϕ_1 for ϕ.

4.5.2 FOUNDATIONS ON WEAKER CLAY LAYER UNDERLAIN BY STRONG SAND LAYER ($\phi_1 = 0$, $\phi_2 = 0$)

Vesic[12] proposed that the ultimate bearing capacity of a foundation supported by a weaker clay layer ($\phi_1 = 0$) underlain by a stronger clay layer ($\phi_2 = 0$) can be expressed as

$$q_u = c_1 m N_c + \gamma_1 D_f \tag{4.65}$$

where

$$N_c = \begin{cases} 5.14 \text{ (for strip foundation)} \\ 6.17 \text{ (for square or circular foundation)} \end{cases}$$

$$m = f\left(\frac{c_1}{c_2}, \frac{H}{B}, \text{ and } \frac{B}{L}\right)$$

Tables 4.2 and 4.3 give the variation of m for strip and square and circular foundations.

TABLE 4.2
Variation of m [Equation (4.65)] for Strip Foundation ($B/L \leq 0.2$)

c_1/c_2	≥ 0.5	0.25	0.167	0.125	0.1
			H/B		
1	1	1	1	1	1
0.667	1	1.033	1.064	1.088	1.109
0.5	1	1.056	1.107	1.152	1.193
0.333	1	1.088	1.167	1.241	1.311
0.25	1	1.107	1.208	1.302	1.389
0.2	1	1.121	1.235	1.342	1.444
0.1	1	1.154	1.302	1.446	1.584

TABLE 4.3

Variation of m [Equation (4.65)] for Strip Foundation ($B/L = 1$)

c_1/c_2	≥ 0.25	0.125	0.083	0.063	0.05
			H/B		
1	1	1	1	1	1
0.667	1	1.028	1.052	1.075	1.096
0.5	1	1.047	1.091	1.131	1.167
0.333	1	1.075	1.143	1.207	1.267
0.25	1	1.091	1.177	1.256	1.334
0.2	1	1.102	1.199	1.292	1.379
0.1	1	1.128	1.254	1.376	1.494

EXAMPLE 4.5

A shallow square foundation 2 m × 2 m in plan is located over a weaker sand layer underlain by a stronger sand layer. Referring to Figure 4.25, given: $D_f = 0.8$ m; $H = 0.5$ m; $\gamma_1 = 16.5$ kN/m³; $\phi_1 = 35°$; $c_1 = 0$; $\gamma_2 = 18.5$ kN/m³; $\phi_2 = 45°$; $c_2 = 0$. Use equation (4.62) and determine the ultimate bearing capacity q_u.

Solution

$H = 0.5$ m; $\phi_1 = 35°$; $\phi_2 = 45°$. From Figures 4.2 and 4.3 for $\phi_1 = 35°$,

$$\frac{D_{(\gamma)}}{B} = 1.0; \quad \frac{D_{(q)}}{B} = 1.9$$

So, $D_{(\gamma)} = 2.0$ m and $D_{(q)} = 3.8$ m. From Table 2.3 for $\phi_1 = 35°$ and $\phi_2 = 45°$, $N_{q(1)} = 33.30$, $N_{q(2)} = 134.88$, and $N_{\gamma(1)} = 37.1$, $N_{\gamma(2)} = 262.7$. Using equations (4.63) and (4.64),

$$N_{\gamma(m)} = 262.7 - \left[\frac{0.5}{2}\right][262.7 - 37.1] = 206.3$$

$$N_{q(m)} = 134.88 - \left[\frac{0.5}{3.8}\right][134.88 - 33.3] = 121.5$$

From equation (4.62),

$$q_u = \tfrac{1}{2}\gamma_1 \lambda_{\gamma s}^* N_{\gamma(m)} + \gamma_1 D_f \lambda_{qs}^* N_{q(m)}$$

From equations (4.4) and (4.5) (*Note*: $H/B = 0.5/2 = 0.25$, and $\phi_1 = 35°$),

$$\lambda_{qs}^* = 1 - m_1\left(\frac{B}{L}\right) \approx 1 - 0.73\left(\frac{2}{2}\right) = 0.27$$

and

$$\lambda_{\gamma s}^* = 1 - m_2\left(\frac{B}{L}\right) = 1 - 0.72\left(\frac{2}{2}\right) = 0.28$$

So,

$$q_u = (0.5)(16.5)(2)(0.28)(206.3) + (16.5)(0.8)(0.27)(121.5) = 953.1 + 433 \approx 1386 \ \text{kN/m}^2$$

CHECK

$$q_u = q_b = \tfrac{1}{2}\gamma_2\lambda_{\gamma s(2)}N_{\gamma(2)} + \gamma_2 D_f\lambda_{qs(2)}N_{q(2)}$$

$$\lambda_{qs(2)} = \lambda_{\gamma s(2)} = 1 + 0.1\left(\frac{B}{L}\right)\tan^2\left(45 + \frac{\phi_2}{2}\right) = 1 + (0.1)\left(\frac{2}{2}\right)\tan^2\left(45 + \frac{45}{2}\right) = 1.583$$

$$q_u = (0.5)(18.5)(2)(1.583)(262.7) + (18.5)(0.8)(1.583)(134.88) = 7693.3 + 3160 \approx \mathbf{10{,}853 \ kN/m^2}$$

So, $q_M = \mathbf{1386 \ kN/m^2}$

4.6 CONTINUOUS FOUNDATION ON WEAK CLAY WITH A GRANULAR TRENCH

In practice, there are several techniques to improve the load-bearing capacity and settlement of shallow foundations on weak compressible soil layers. One of those techniques is the use of a granular trench under a foundation. Figure 4.26 shows a continuous rough foundation on a granular trench made in weak soil extending to a great depth. The width of the trench is W, the width of the foundation is B, and the depth of the trench is H. The width W of the trench can be smaller or larger than B. The parameters of the stronger trench material and the weak soil for bearing capacity calculation are as follows:

	Trench Material	Weak Soil
Angle of friction	ϕ_1	ϕ_2
Cohesion	c_1	c_2
Unit weight	γ_1	γ_2

Madhav and Vitkar[23] assumed a general shear failure mechanism in the soil under the foundation to analyze the ultimate bearing capacity of the foundation using the upper bound limit analysis suggested by Drucker and Prager,[24] and this is shown in Figure 4.26. The failure zone in the soil can be divided into subzones, and they are as follows:

1. An active Rankine zone ABC with a wedge angle of ξ
2. A mixed transition zone such as BCD bounded by angle θ_1. CD is an arc of a log spiral defined by the equation

$$r = r_0 e^{\theta \tan\phi_1}$$

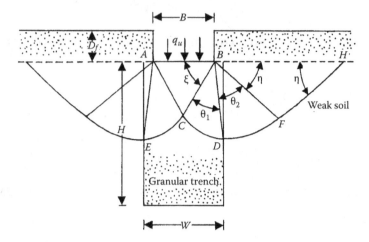

FIGURE 4.26 Continuous rough foundation on weak soil with a granular trench.

where

ϕ_1 = angle of friction of the trench material

3. A transition zone such as *BDF* with a central angle θ_2. *DF* is an arc of a log spiral defined by the equation

$$r = r_0 e^{\theta \tan \phi_2}$$

4. A Rankine passive zone like *BFH*.

Note that θ_1 and θ_2 are functions of ξ, η, W/B, and ϕ_1.

By using the upper bound limit analysis theorem, Madhav and Vitkar[23] expressed the ultimate bearing capacity of the foundation as

$$q_u = c_2 N_{c(T)} + D_f \gamma_2 N_{q(T)} + \left(\frac{\gamma_2 B}{2}\right) N_{\gamma(T)} \qquad (4.66)$$

where

$N_{c(T)}, N_{q(T)}, N_{\gamma(T)}$ = bearing capacity factors with the presence of the trench

The variations of the bearing capacity factors [that is, $N_{c(T)}, N_{q(T)}$, and $N_{\gamma(T)}$] for purely granular trench soil ($c_1 = 0$) and soft saturated clay (with $\phi_2 = 0$ and $c_2 = c_u$) determined by Madhav and Vitkar[23] are given in Figures 4.27, 4.28, and 4.29. The values of $N_{\gamma(T)}$ given in Figure 4.29 are for $\gamma_1/\gamma_2 = 1$. In an actual case, the ratio γ_1/γ_2 may be different than one; however, the error for this assumption is less than 10%.

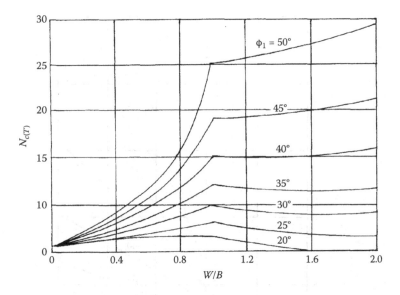

FIGURE 4.27 Madhav and Vitkar's bearing capacity factor $N_{c(T)}$.

Sufficient experimental results are not available in the literature to verify the above theory. Hamed, Das, and Echelberger[25] conducted several laboratory model tests to determine the variation of the ultimate bearing capacity of a strip foundation resting on a granular trench (sand; $c_1 = 0$) made in a saturated soft clay medium ($\phi_2 = 0$; $c_2 = c_u$).

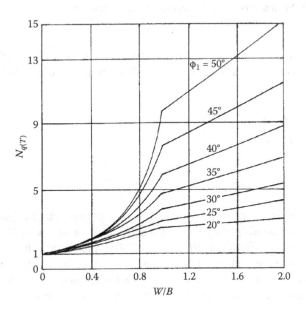

FIGURE 4.28 Madhav and Vitkar's bearing capacity factor $N_{q(T)}$.

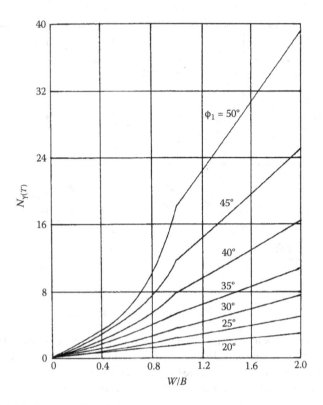

FIGURE 4.29 Madhav and Vitkar's bearing capacity factor $N_{\gamma(T)}$.

For these tests the width of the foundation B was kept equal to the width of the trench W, and the ratio of H/B was varied. The details of the tests are as follows:

Series I:
 $\phi_1 = 40°$, $c_1 = 0$
 $\phi_2 = 0$, $c_2 = c_u = 1656$ kN/m²
Series II:
 $\phi_1 = 43°$, $c_1 = 0$
 $\phi_2 = 0$, $c_2 = c_u = 1656$ kN/m²

For both test series D_f was kept equal to zero (that is, surface foundation). For each test series the ultimate bearing capacity q_u increased with H/B almost linearly, reaching a maximum at $H/B \approx 2.5$ to 3. The maximum values of q_u obtained experimentally were compared with those presented by Madhav and Vitkar.[23] The theoretical values were about 40%–70% higher than those obtained experimentally. Further refinement to the theory is necessary to provide more realistic results.

4.7 SHALLOW FOUNDATION ABOVE A VOID

Mining operations may leave underground voids at relatively shallow depths. Additionally, in some instances, void spaces occur when soluble bedrock dissolves at the interface of the soil and bedrock. Estimating the ultimate bearing capacity of shallow foundations constructed over these voids, as well as the stability of the foundations, is gradually becoming an important issue. Only a few studies have been published so far. Baus and Wang[26] reported some experimental results for the ultimate bearing capacity of a shallow rough continuous foundation located above voids as shown in Figure 4.30. It is assumed that the top of the rectangular void is located at a depth H below the bottom of the foundation. The void is continuous and has cross-sectional dimensions of $W' \times H'$. The laboratory tests of Baus and Wang[26] were conducted with soil having the following properties:

Friction angle of soil $\phi = 13.5°$
Cohesion $= 65.6$ kN/m^2
Modulus in compression $= 4670$ kN/m^2
Modulus in tension $= 10,380$ kN/m^2
Poisson's ratio $= 0.28$
Unit weight of compacted soil $\gamma = 18.42$ kN/m^3

The results of Baus and Wang[26] are shown in a nondimensional form in Figure 4.31. Note that the results of the tests that constitute Figure 4.31 are for the case of $D_f = 0$. From this figure the following conclusions can be drawn:

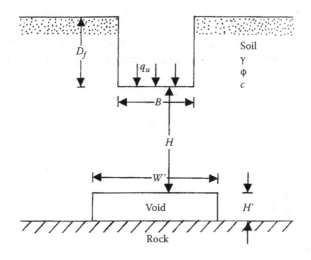

FIGURE 4.30 Shallow continuous rough foundation over a void.

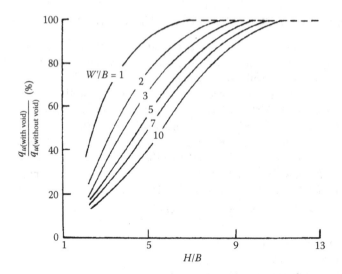

FIGURE 4.31 Experimental bearing capacity of a continuous foundation as a function of void size and location. *Source:* From Baus, R.L., and M.C. Wang. 1983. Bearing capacity of strip booking above void. *J. Geotech. Eng.*, ASCE, 109(1):1.

1. For a given H/B, the ultimate bearing capacity decreases with the increase in the void width, W'.
2. For any given W'/B, there is a critical H/B ratio beyond which the void has no effect on the ultimate bearing capacity. For $W'/B = 10$, the value of the critical H/B is about 12.

Baus and Wang[26] conducted finite analysis to compare the validity of their experimental findings. In the finite element analysis, the soil was treated as an *elastic–perfectly plastic material*. They also assumed that Hooke's law is valid in the elastic range and that the soil follows the von Mises yield criterion in the perfectly plastic range, or

$$f = \alpha J_1 + \sqrt{J_2} = k' \tag{4.67}$$

$$\dot{f} = 0 \tag{4.68}$$

where
f = yield function

$$\alpha = \frac{\tan \phi}{(9 + 12 \tan \phi)^{0.5}} \tag{4.69}$$

$$k' = \frac{3c}{(9 + 12 \tan \phi)^{0.5}} \tag{4.70}$$

J_1 = first stress invariant
J_2 = second stress invariant

The relationships shown in equations (4.69) and (4.70) are based on the study of Drucker and Prager.[24] The results of the finite element analysis have shown good agreement with experiments.

4.8 FOUNDATION ON A SLOPE

In 1957 Meyerhof[27] proposed a theoretical solution to determine the ultimate bearing capacity of a shallow foundation located on the face of a slope. Figure 4.32 shows the nature of the plastic zone developed in the soil under a rough continuous foundation (width $= B$) located on the face of a slope. In Figure 4.32, abc is the elastic zone, acd is a radial shear zone, and ade is a mixed shear zone. The normal and shear stresses on plane ae are p_o and s_o, respectively. Note that the slope makes an angle β with the horizontal. The shear strength parameters of the soil are c and ϕ, and its unit weight is equal to γ. As in equation (2.71), the ultimate bearing capacity can be expressed as

$$q_u = cN_c + p_oN_c + \tfrac{1}{2}\gamma BN_\gamma \qquad (4.71)$$

The preceding relationship can also be expressed as

$$q_u = cN_{cq} + \tfrac{1}{2}\gamma BN_{\gamma q} \qquad (4.72)$$

where
$N_{cq}, N_{\gamma q} =$ bearing capacity factors

For purely cohesive soil (that is, $\phi = 0$),

$$q_u = cN_{cq} \qquad (4.73)$$

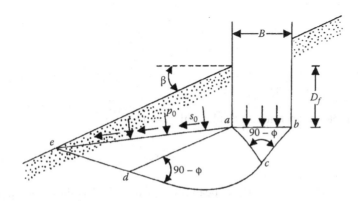

FIGURE 4.32 Nature of plastic zone under a rough continuous foundation on the face of a slope.

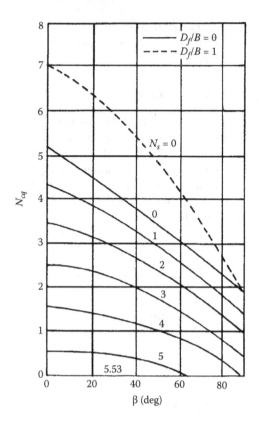

FIGURE 4.33 Variation of Meyerhof's bearing capacity factor N_{cq} for a purely cohesive soil (foundation on a slope).

Figure 4.33 shows the variation of N_{cq} with slope angle β and the *slope stability number N_s*. Note that

$$N_s = \frac{\gamma H}{c} \qquad (4.74)$$

where
H = height of the slope

In a similar manner, for a granular soil ($c = 0$),

$$q_u = \tfrac{1}{2}\gamma B N_{\gamma q} \qquad (4.75)$$

The variation of $N_{\gamma q}$ (for $c = 0$) applicable to equation (4.75) is shown in Figure 4.34.

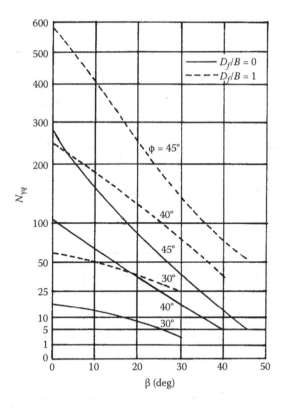

FIGURE 4.34 Variation of Meyerhof's bearing capacity factor $N_{\gamma q}$ for a purely granular soil (foundation on a slope).

4.9 FOUNDATION ON TOP OF A SLOPE

4.9.1 MEYERHOF'S SOLUTION

Figure 4.35 shows a rough continuous foundation of width B located on top of a slope of height H. It is located at a distance b from the edge of the slope. The ultimate bearing capacity of the foundation can be expressed by equation (4.72), or

$$q_u = cN_{cq} + \tfrac{1}{2}\gamma B N_{\gamma q} \qquad (4.76)$$

Meyehof[27] developed the theoretical variations of N_{cq} for a purely cohesive soil ($\phi = 0$) and $N_{\gamma q}$ for a granular soil ($c = 0$), and these variations are shown in Figures 4.36 and 4.37. Note that, for purely cohesive soil (Figure 4.36),

$$q_u = cN_{cq}$$

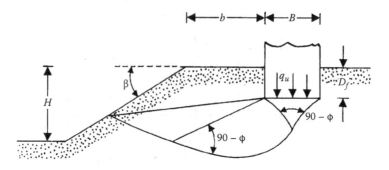

FIGURE 4.35 Continuous foundation on a slope.

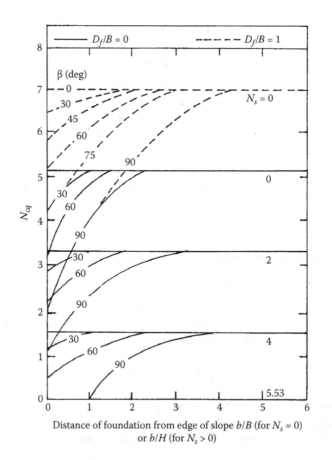

Distance of foundation from edge of slope b/B (for $N_s = 0$)
or b/H (for $N_s > 0$)

FIGURE 4.36 Meyerhof's bearing capacity factor N_{cq} for a purely cohesive soil (foundation on top of a slope).

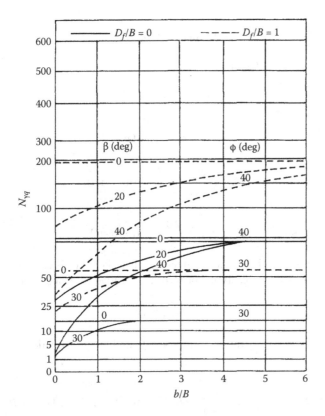

FIGURE 4.37 Meyerhof's bearing capacity factor $N_{\gamma q}$ for a granular soil (foundation on top of a slope).

and for granular soil (Figure 4.37),

$$q_u = \tfrac{1}{2}\gamma B N_{\gamma q}$$

It is important to note that, when using Figure 4.36, the stability number N_s should be taken as zero when $B < H$. If $B \geq H$, the curve for the actual stability number should be used.

4.9.2 SOLUTIONS OF HANSEN AND VESIC

Referring to the condition of $b = 0$ in Figure 4.35 (that is, the foundation is located at the edge of the slope), Hansen[28] proposed the following relationship for the ultimate bearing capacity of a continuous foundation:

$$q_u = cN_c\lambda_{c\beta} + qN_q\lambda_{q\beta} + \tfrac{1}{2}\gamma B N_\gamma \lambda_{\gamma\beta} \tag{4.77}$$

where

N_c, N_q, N_γ = bearing capacity factors (see Table 2.3 for N_c and N_q and Table 2.4 for N_γ)

$\lambda_{c\beta}$, $\lambda_{q\beta}$, $\lambda_{\gamma\beta}$ = slope factors

$q = \gamma D_f$

According to Hansen,[28]

$$\lambda_{q\beta} = \lambda_{\gamma\beta} = (1 - \tan \beta)^2 \tag{4.78}$$

$$\lambda_{c\beta} = \frac{N_q \lambda_{q\beta} - 1}{N_q - 1} \quad \text{(for } \phi > 0) \tag{4.79}$$

$$\lambda_{c\beta} = 1 - \frac{2\beta}{\pi + 2} \quad \text{(for } \phi = 0) \tag{4.80}$$

For the $\phi = 0$ condition Vesic[12] pointed out that, with the absence of weight due to the slope, the bearing capacity factor N_γ has a negative value and can be given as

$$N_\gamma = -2 \sin \beta \tag{4.81}$$

Thus, for the $\phi = 0$ condition with $N_c = 5.14$ and $N_q = 1$, equation (4.77) takes the form

$$q_u = c(5.14)\left(1 - \frac{2\beta}{5.14}\right) + \gamma D_f(1 - \tan \beta)^2 - \gamma B \sin \beta(1 - \tan \beta)^2$$

or

$$q_u = (5.14 - 2\beta)c + \gamma D_f(1 - \tan \beta)^2 - \gamma B \sin \beta(1 - \tan \beta)^2 \tag{4.82}$$

4.9.3 SOLUTION BY LIMIT EQUILIBRIUM AND LIMIT ANALYSIS

Saran, Sud, and Handa[29] provided a solution to determine the ultimate bearing capacity of shallow continuous foundations *on the top of a slope* (Figure 4.35) using the limit equilibrium and limit analysis approach. According to this theory, for a strip foundation,

$$q_u = cN_c + qN_q + \tfrac{1}{2}\gamma BN_\gamma \tag{4.83}$$

where

N_c, N_q, N_γ = bearing capacity factors

$q = \gamma D_f$

Referring to the notations used in Figure 4.35, the numerical values of N_c, N_q, and N_γ are given in Table 4.4.

TABLE 4.4

Bearing Capacity Factors Based on Saran, Sud, and Handa's Analysis

Factor	β (deg)	$\dfrac{D_f}{B}$	$\dfrac{b}{B}$	40	35	30	25	20	15	10
						Soil Friction Angle ϕ (deg)				
N_γ	30	0	0	25.37	12.41	6.14	3.20	1.26	0.70	0.10
	20			53.48	24.54	11.62	5.61	4.27	1.79	0.45
	10			101.74	43.35	19.65	9.19	4.35	1.96	0.77
	0			165.39	66.59	28.98	13.12	6.05	2.74	1.14
	30	0	1	60.06	34.03	18.95	10.33	5.45	0.00	—
	20			85.98	42.49	21.93	11.42	5.89	1.35	—
	10			125.32	55.15	25.86	12.26	6.05	2.74	—
	0			165.39	66.59	28.89	13.12	6.05	2.74	—
	30	1	0	91.87	49.43	26.39	—	—	—	—
	25			115.65	59.12	28.80	—	—	—	—
	20			143.77	66.00	28.89	—	—	—	—
	≤15			165.39	66.59	28.89	—	—	—	—
	30	1	1	131.34	64.37	28.89	—	—	—	—
	25			151.37	66.59	28.89	—	—	—	—
	≤20			166.39	66.59	28.89	—	—	—	—
N_q	30	1	0	12.13	16.42	8.98	7.04	5.00	3.60	—
	20			12.67	19.48	16.80	12.70	7.40	4.40	—
	≤10			81.30	41.40	22.50	12.70	7.40	4.40	—
	30	1	1	28.31	24.14	22.5	—	—	—	—
	20			42.25	41.4	22.5	—	—	—	—
	≤10			81.30	41.4	22.5	—	—	—	—
N_c	50	0	0	21.68	16.52	12.60	10.00	8.60	7.10	5.50
	40			31.80	22.44	16.64	12.80	10.04	8.00	6.25
	30			44.80	28.72	22.00	16.20	12.20	8.60	6.70
	20			63.20	41.20	28.32	20.60	15.00	11.30	8.76
	≤10			88.96	55.36	36.50	24.72	17.36	12.61	9.44
	50	0	1	38.80	30.40	24.20	19.70	16.42	—	—
	40			48.00	35.40	27.42	21.52	17.28	—	—
	30			59.64	41.07	30.92	23.60	17.36	—	—
	20			75.12	50.00	35.16	27.72	17.36	—	—
	≤10			95.20	57.25	36.69	24.72	17.36	—	—
	50	1	0	35.97	28.11	22.38	18.38	15.66	10.00	—
	40			51.16	37.95	29.42	22.75	17.32	12.16	—
	30			70.59	50.37	36.20	24.72	17.36	12.16	—
	20			93.79	57.20	36.20	24.72	17.36	12.16	—
	≤10			95.20	57.20	36.20	24.72	17.36	12.16	—
	50	1	1	53.65	42.47	35.00	24.72	—	—	—
	40			67.98	51.61	36.69	24.72	—	—	—
	30			85.38	57.25	36.69	24.72	—	—	—
	≤20			95.20	57.25	36.69	24.72	—	—	—

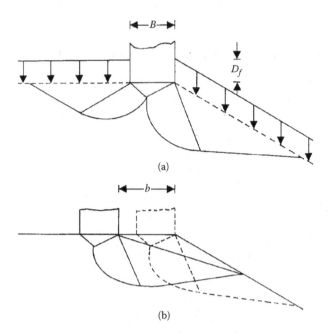

FIGURE 4.38 Schematic diagram of failure zones for embedment and setback: (a) $D_f/B >$ 0; (b) $b/B > 0$.

4.9.4 STRESS CHARACTERISTICS SOLUTION

As shown in equation (4.76), for granular soils (that is, $c = 0$)

$$q_u = \tfrac{1}{2}\gamma B N_{\gamma q} \qquad (4.84)$$

Graham, Andrews, and Shields[30] provided a solution for the bearing capacity factor $N_{\gamma q}$ for a shallow continuous foundation on the top of a slope in *granular soil* based on the method of stress characteristics. Figure 4.38 shows the schematics of the failure zone in the soil for embedment (D_f/B) and setback (b/B) assumed for this analysis. The variations of $N_{\gamma q}$ obtained by this method are shown in Figures 4.39, 4.40, and 4.41.

EXAMPLE 4.6

Refer to Figure 4.35 and consider a continuous foundation on a saturated clay slope. Given, for the slope: $H = 7$ m; $\beta = 30°$; $\gamma = 18.5$ kN/m³; $\phi = 0$, $c = 49$ kN/m²; and given, for the foundation: $D_f = 1.5$ m; $B = 1.5$ m; $b = 0$. Estimate the ultimate bearing capacity by:

a. Meyerhof's method [equation (4.76)]
b. Hansen and Vesic's method [equation (4.82)]

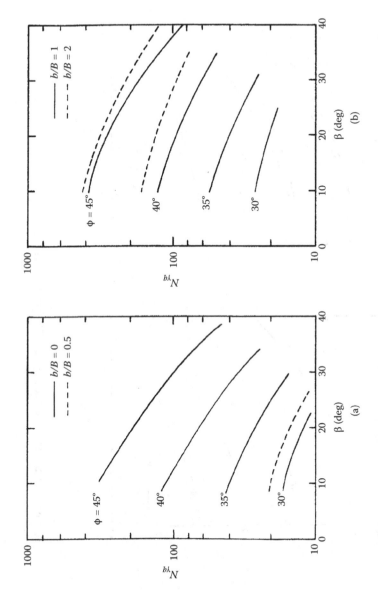

FIGURE 4.39 Graham, Andrews, and Shields' theoretical values of $N_{\gamma q}$ ($D_f/B = 0$).

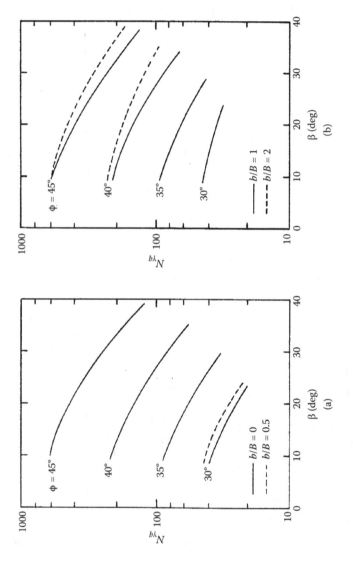

FIGURE 4.40 Graham, Andrews, and Shields' theoretical values of $N_{\gamma q}$ ($D_f/B = 0.5$).

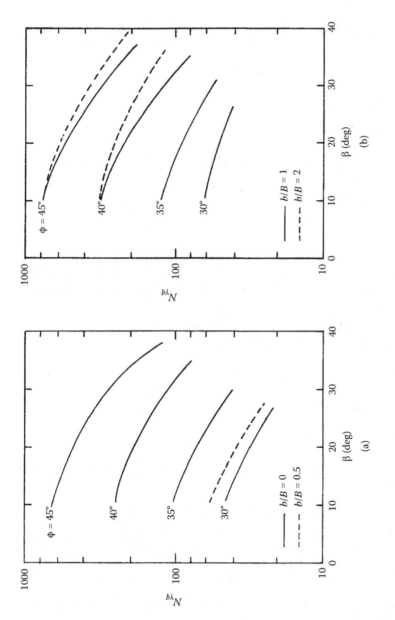

FIGURE 4.41 Graham, Andrews, and Shields' theoretical values of $N_{\gamma q}$ ($D_f/B = 1$).

Solution

Part a:

$$q_{qu} = cN_{cq}$$

Given $D_f/B = 1.5/1.5 = 1$; $b/B = 0/1.5 = 0$. Since $H/B > 1$, use $N_s = 0$.
From Figure 4.36, for $D_f/B = 1$; $b/B = 0$, $\beta = 30°$, and $N_s = 0$, the value of N_{cq} is about 5.85. So,

$$q_u = (49)(5.85) = \textbf{286.7 kN/m}^2$$

Part b:
From equation (4.82),

$$q_u = (5.14 - 2\beta)c + \gamma D_f(1 - \tan \beta)^2 - \gamma B \sin \beta(1 - \tan \beta)^2$$

$$= \left[5.14 - (2)\left(\frac{\pi}{80} \times 30 \right) \right](49) + (18.5)(1.5)(1 - \tan 30)^2 - (18.5)(1.5)(\sin 30)(1 - \tan 30)^2$$

$$= \textbf{203 kN/m}^2$$

EXAMPLE 4.7

Refer to Figure 4.35 and consider a continuous foundation on a slope of granular soil. Given, for the slope: $H = 6$ m; $\beta = 30°$; $\gamma = 16.8$ kN/m^3; $\phi = 40°$; $c = 0$; and given, for the foundation: $D_f = 1.5$ m; $B = 1.5$ m; $b = 1.5$ m. Estimate the ultimate bearing capacity by:

a. Meyerhof's method [equation (4.76)]
b. Saran, Sud, and Handa's method [equation (4.83)]
c. The stress characteristic solution [equation (4.84)]

Solution

Part a:
For granular soil ($c = 0$), from equation (4.76),

$$q_u = \tfrac{1}{2}\gamma B N_{\gamma q}$$

Given: $b/B = 1.5/1.5 = 1$; $D_f/B = 1.5/1.5 = 1$; $\phi = 40°$; and $\beta = 30°$. From Figure 4.37, $N_{\gamma q} \approx 120$. So,

$$q_u = \tfrac{1}{2}(16.8)(1.5)(120) = \textbf{1512 kN/m}^2$$

Part b:
For $c = 0$, from equation (4.83),

$$q_u = qN_q + \tfrac{1}{2}\gamma B N_\gamma$$

For $b/B = 1$: $D_f/B = 1$; $\phi = 40°$; and $\beta = 30°$. The value of $N_\gamma = 131.34$ and the value of $N_q = 28.31$ (Table 4.4).

$$q_u = (16.8)(1.5)(28.31) + \tfrac{1}{2}(16.8)(1.5)(131.34) \approx \textbf{2368 kN/m}^2$$

Part c:

From equation (4.84),

$$q_u = \tfrac{1}{2}\gamma B N_{\gamma q}$$

From Figure 4.41b, $N_{\gamma q} \approx 110$,

$$q_u = \tfrac{1}{2}(16.8)(1.5)(110) = \mathbf{1386\ kN/m^2}$$

REFERENCES

1. Prandtl, L. 1921. Uber die eindringungsfestigkeit plastisher baustoffe und die festigkeit von schneiden. *Z. Ang. Math. Mech.* 1(1): 15.
2. Reissner, H. 1924. Zum erddruckproblem, in *Proc., I Intl. Conf. Appl. Mech.*, Delft, The Netherlands, 295.
3. Lundgren, H., and K. Mortensen. 1953. Determination by the theory of plasticity of the bearing capacity of continuous footings on sand, in *Proc., III Int. Conf. Soil Mech. Found. Eng.*, Zurich, Switzerland, 1: 409.
4. Mandel, J., and J. Salencon. 1972. Force portante d'un sol sur une assise rigide (étude theorizue). *Geotechnique* 22(1): 79.
5. Meyerhof, G. G., and T. K. Chaplin. 1953. The compression and bearing capacity of cohesive soils. *Br. J. Appl. Phys.* 4: 20.
6. Meyerhof, G. G. 1974. Ultimate bearing capacity of footings on sand layer overlying clay. *Canadian Geotech. J.* 11(2): 224.
7. Milovic, D. M., and J. P. Tournier. 1971. Comportement de foundations reposant sur une coche compressible d'épaisseur limitée, in *Proc., Conf. Comportement des Sols Avant la Rupture,* Paris, France: 303.
8. Pfeifle, T. W., and B. M. Das. 1979. Bearing capacity of surface footings on sand layer resting on rigid rough base. *Soils and Foundations* 19(1): 1.
9. Cerato, A. B., and A. J. Lutenegger. 2006. Bearing capacity of square and circular footings on a finite layer of granular soil underlain by a rigid base. *J. Geotech. Geoenv. Eng.*, ASCE, 132(11): 1496.
10. Mandel, J., and J. Salencon. 1969. Force portante d'un sol sur une assise rigide, in *Proc., VII Int. Conf. Soil Mech. Found Eng.*, Mexico City, 2: 157.
11. Buisman, A. S. K. 1940. *Grondmechanica.* Delft: Waltman.
12. Vesic, A. S. 1975. Bearing capacity of shallow foundations, in *Foundation engineering handbook,* ed. H. F. Winterkorn and H. Y. Fang, 121. New York: Van Nostrand Reinhold Co.
13. DeBeer, E. E. 1975. Analysis of shallow foundations, in *Geotechnical modeling and applications*, ed. S. M. Sayed, 212. Gulf Publishing Co. Houston, USA.
14. Casagrande, A., and N. Carrillo. 1954. Shear failure in anisotropic materials, in *Contribution to soil mechanics 1941–53*, Boston Society of Civil Engineers, 122.
15. Reddy, A. S., and R. J. Srinivasan. 1967. Bearing capacity of footings on layered clays. *J. Soil Mech. Found. Div.*, ASCE, 93(SM2): 83.
16. Lo, K. Y. 1965. Stability of slopes in anisotropic soil. *J. Soil Mech. Found. Div.*, ASCE, 91(SM4): 85.
17. Meyerhof, G. G., and A. M. Hanna. 1978. Ultimate bearing capacity of foundations on layered soils under inclined load. *Canadian Geotech. J.* 15(4): 565.
18. Caquot, A., and J. Kerisel. 1949. *Tables for the calculation of passive pressure, active pressure, and bearing capacity of foundations.* Paris: Gauthier-Villars.

19. Meyerhof, G. G. 1963. Some recent research on the bearing capacity of foundations. *Canadian Geotech. J.* 1(1): 16.

20. Hanna, A. M., and G. G. Meyerhof. 1980. Design charts for ultimate bearing capacity for sands overlying clays. *Canadian Geotech. J.* 17(2): 300.

21. Hanna, A. M. 1981. Foundations on strong sand overlying weak sand. *J. Geotech. Eng,,* ASCE, 107(GT7): 915.

22. Hanna, A. M. 1982. Bearing capacity of foundations on a weak sand layer overlying a strong deposit. *Canadian Geotech. J.* 19(3): 392.

23. Madhav, M. R., and P. P. Vitkar. 1978. Strip footing on weak clay stabilized with a granular trench or pile. *Canadian Geotech. J.* 15(4): 605.

24. Drucker, D. C., and W. Prager. 1952. Soil mechanics and plastic analysis of limit design. *Q. Appl. Math.* 10: 157.

25. Hamed, J. T., B. M. Das, and W. F. Echelberger. 1986. Bearing capacity of a strip foundation on granular trench in soft clay. *Civil Engineering for Practicing and Design Engineers* 5(5): 359.

26. Baus, R. L., and M. C. Wang. 1983. Bearing capacity of strip footing above void. *J. Geotech. Eng.,* ASCE, 109(GT1): 1.

27. Meyerhof, G. G. 1957. The ultimate bearing capacity of foundations on slopes, in *Proc., IV Int. Conf. Soil Mech. Found. Eng.*, London, England, 1: 384.

28. Hansen, J. B. 1970. *A revised and extended formula for bearing capacity*, Bulletin 28. Copenhagen: Danish Geotechnical Institute.

29. Saran, S., V. K. Sud, and S. C. Handa. 1989. Bearing capacity of footings adjacent to slopes. *J. Geotech. Eng.,* ASCE, 115(4): 553.

30. Graham, J., M. Andrews, and D. H. Shields. 1988. Stress characteristics for shallow footings in cohesionless slopes. *Canadian Geotech. J.* 25(2): 238.

5 Settlement and Allowable Bearing Capacity

5.1 INTRODUCTION

Various theories relating to the ultimate bearing capacity of shallow foundations were presented in Chapters 2, 3, and 4. In section 2.12 a number of definitions for the allowable bearing capacity were discussed. In the design of any foundation, one must consider the safety against *bearing capacity failure* as well as against *excessive settlement* of the foundation. In the design of most foundations, there are specifications for allowable levels of settlement. Refer to Figure 5.1, which is a plot of load per unit area q versus settlement S for a foundation. The ultimate bearing capacity is realized at a settlement level of S_u. Let S_{all} be the allowable level of settlement for the foundation and $q_{all(S)}$ be the corresponding allowable bearing capacity. If FS is the factor of safety against bearing capacity failure, then the allowable bearing capacity is $q_{all(b)} = q_u/FS$. The settlement corresponding to $q_{all(b)}$ is S'. For foundations with smaller widths of B, S' may be less than S_{all}; however, $S_{all} < S'$ for larger values of B. Hence, for smaller foundation widths, the bearing capacity controls; for larger foundation widths, the allowable settlement controls. This chapter describes the procedures for estimating the settlements of foundations under load and thus the allowable bearing capacity.

The settlement of a foundation can have three components: (a) elastic settlement S_e, (b) primary consolidation settlement S_c, and (c) secondary consolidation settlement S_s.

The total settlement S_t can be expressed as

$$S_t = S_e + S_c + S_s$$

For any given foundation, one or more of the components may be zero or negligible.

Elastic settlement is caused by deformation of dry soil, as well as moist and saturated soils, without any change in moisture content. Primary consolidation settlement is a time-dependent process that occurs in clayey soils located below the groundwater table as a result of the volume change in soil because of the expulsion of water that occupies the void spaces. Secondary consolidation settlement follows the primary consolidation process in saturated clayey soils and is a result of the plastic adjustment of soil fabrics. The procedures for estimating the above three types of settlements are discussed in this chapter.

Any type of settlement is a function of the additional stress imposed on the soil by the foundation. Hence, it is desirable to know the relationships for calculating the stress increase in the soil caused by application of load to the foundation. These

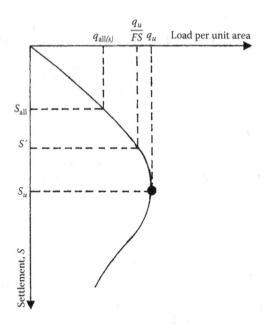

FIGURE 5.1 Load–settlement curve for shallow foundation.

relationships are given in section 5.2 and are derived assuming that the soil is a semi-infinite, elastic, and homogeneous medium.

5.2   STRESS INCREASE IN SOIL DUE TO APPLIED LOAD—BOUSSINESQ'S SOLUTION

5.2.1   POINT LOAD

Boussinesq[1] developed a mathematic relationship for the stress increase due to a point load Q acting on the surface of a semi-infinite mass. In Figure 5.2 the stress increase at a point A is shown in the Cartesian coordinate system, and the stress increase in the cylindrical coordinate system is shown in Figure 5.3. The components of the stress increase can be given by the following relationships:

Cartesian Coordinate System (Figure 5.2)

$$\sigma_z = \frac{3Qz^3}{2\pi R^5} \tag{5.1}$$

$$\sigma_x = \frac{3Q}{2\pi}\left\{\frac{x^2 z}{R^5} + \frac{1-2v}{3}\left[\frac{1}{R(R+z)} - \frac{(2R+z)x^2}{R^3(R+z)^2} - \frac{z}{R^3}\right]\right\} \tag{5.2}$$

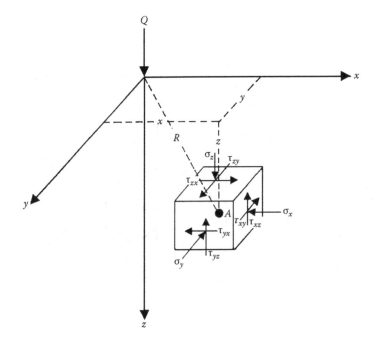

FIGURE 5.2 Boussinesq's problem—stress increase at a point in the Cartesian coordinate system due to a point load on the surface.

$$\sigma_y = \frac{3Q}{2\pi}\left\{\frac{y^2 z}{R^5} + \frac{1-2v}{3}\left[\frac{1}{R(R+z)} - \frac{(2R+z)y^2}{R^3(R+z)^2} - \frac{z}{R^3}\right]\right\} \tag{5.3}$$

$$\tau_{xy} = \frac{3Q}{2\pi}\left[\frac{xyz}{R^5} - \frac{1-2v}{3}\frac{(2R+z)xy}{R^3(R+z)^2}\right] \tag{5.4}$$

$$\tau_{xz} = \frac{3Q}{2\pi}\frac{xz^2}{R^5} \tag{5.5}$$

$$\tau_{yz} = \frac{3Q}{2\pi}\frac{yz^2}{R^5} \tag{5.6}$$

where
  σ = normal stress
  τ = shear stress
  $R = \sqrt{z^2 + r^2}$
  $r = \sqrt{x^2 + y^2}$
  v = Poisson's ratio

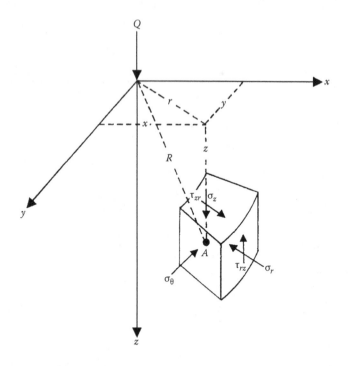

FIGURE 5.3 Boussinesq's problem—stress increase at a point in the cylindrical coordinate system due to a point load on the surface.

Cylindrical Coordinate System (Figure 5.3)

$$\sigma_z = \frac{3Qz^3}{2\pi R^5} \tag{5.7}$$

$$\sigma_r = \frac{Q}{2\pi}\left[\frac{3zr^2}{R^5} - \frac{1-2v}{R(R+z)}\right] \tag{5.8}$$

$$\sigma_\theta = \frac{Q}{2\pi}(1-2v)\left[\frac{1}{R(R+z)} - \frac{z}{R^3}\right] \tag{5.9}$$

$$\tau_{rz} = \frac{3Qrz^2}{2\pi R^5} \tag{5.10}$$

5.2.2 UNIFORMLY LOADED FLEXIBLE CIRCULAR AREA

Boussinesq's solution for a point load can be extended to determine the stress increase due to a uniformly loaded *flexible* circular area on the surface of a semi-infinite mass (Figure 5.4). In Figure 5.4, the circular area has a radius R, and the

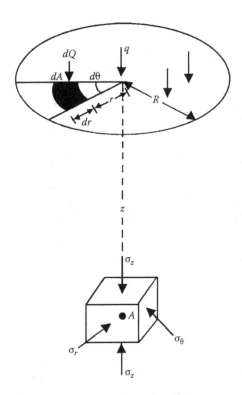

FIGURE 5.4 Stress increase below the center of a uniformly loaded flexible circular area.

uniformly distributed load per unit area is q. If the components of stress increase at a point A *below the center* are to be determined, then we consider an elemental area $dA = rd\theta dr$. The load on the elemental area is $dQ = qrd\theta dr$. This can be treated as a point load. Now the vertical stress increase $d\sigma_z$ at A due to dQ can be obtained by substituting dQ for Q and $\sqrt{r^2 + z^2}$ for R in equation (5.7). Thus,

$$d\sigma_z = \frac{3z^3 qrd\theta dr}{2\pi(r^2 + z^2)^{5/2}}$$

The vertical stress increase due to the entire loaded area σ_z is then

$$\sigma_z = \int d\sigma_z = \int_{r=0}^{R} \int_{\theta=0}^{2\pi} \frac{3z^3 qrd\theta dr}{2\pi(r^2 + z^2)^{5/2}} = q\left[1 - \frac{z^3}{(R^2 + z^2)^{3/2}}\right] \qquad (5.11)$$

Similarly, the magnitudes of σ_θ and σ_r *below the center* can be obtained as

$$\sigma_r = \sigma_\theta = \frac{q}{2}\left[1 + 2v - \frac{2(1+v)z}{(R^2 + z^2)^{1/2}} + \frac{z^3}{(R^2 + z^2)^{3/2}}\right] \qquad (5.12)$$

TABLE 5.1
Variation of σ_z/q at a Point A (Figure 5.5)

z/R	σ_z/q					
	r/R = 0	r/R = 0.2	r/R = 0.4	r/R = 0.6	r/R = 0.8	r/R = 1.0
0.0	1.000	1.000	1.000	1.000	1.000	0.500
0.2	0.992	0.991	0.987	0.970	0.890	0.468
0.4	0.979	0.943	0.920	0.860	0.713	0.435
0.6	0.864	0.852	0.814	0.732	0.591	0.400
0.8	0.756	0.742	0.699	0.619	0.504	0.366
1.0	0.646	0.633	0.591	0.525	0.434	0.332
1.5	0.424	0.416	0.392	0.355	0.308	0.288
2.0	0.284	0.281	0.268	0.248	0.224	0.196
2.5	0.200	0.197	0.196	0.188	0.167	0.151
3.0	0.146	0.145	0.141	0.135	0.127	0.118
4.0	0.087	0.086	0.085	0.082	0.080	0.075
5.0	0.057	0.057	0.056	0.054	0.053	0.052

Table 5.1 gives the variation of σ_z/q at any point A below a circularly loaded flexible area for $r/R = 0$ to 1 (Figure 5.5). A more detailed tabulation of the stress increase (that is, σ_z, σ_θ, σ_r, and τ_{rz}) below a uniformly loaded flexible area is given by Ahlvin and Ulery.[2]

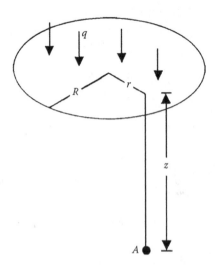

FIGURE 5.5 Stress increase below any point under a uniformly loaded flexible circular area.

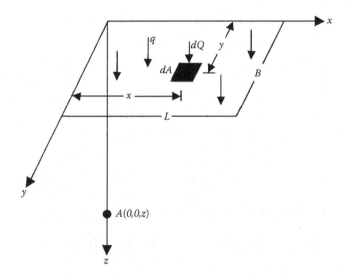

FIGURE 5.6 Stress increase below the corner of a uniformly loaded flexible rectangular area.

5.2.3 Uniformly Loaded Flexible Rectangular Area

Figure 5.6 shows a *flexible* rectangular area of length L and width B subjected to a uniform vertical load of q per unit area. The load on the elemental area dA is equal to $dQ = q\,dx\,dy$. This can be treated as an elemental point load. The vertical stress increase $d\sigma_z$ due to this at A, which is located at a depth z below the corner of the rectangular area, can be obtained by using equation (5.7), or

$$d\sigma_z = \frac{3qz^3dxdy}{2\pi(x^2+y^2+z^2)^{5/2}} \tag{5.13}$$

Hence, the vertical stress increase at A due to the entire loaded area is

$$\sigma_z = \int d\sigma_z = \int_{y=0}^{B}\int_{x=0}^{L} \frac{3qz^3dxdy}{2\pi(r^2+z^2)^{5/2}} = qI \tag{5.14}$$

where

$$I = \frac{1}{4\pi}\left[\frac{2mn(m^2+n^2+1)^{0.5}}{m^2+n^2+m^2n^2+1}\times\frac{m^2+n^2+2}{m^2+n^2+1}+\tan^{-1}\frac{2mn(m^2+n^2+1)^{0.5}}{m^2+n^2-m^2n^2+1}\right] \tag{5.15}$$

$$m = \frac{B}{z}$$

$$n = \frac{L}{z}$$

Table 5.2 shows the variation of I with m and n. The stress below any other point C below the rectangular area (Figure 5.7) can be obtained by dividing it into four rectangles as shown. For rectangular area 1, $m_1 = B_1/z$; $n_1 = L_1/z$. Similarly, for rectangles 2, 3,

TABLE 5.2
Variation of I with m and n

n \ m	0.1	0.2	0.3	0.4	0.5	0.6	0.7	0.8	0.9	1.0	1.2	1.4	1.6	1.8	2.0	2.5	3.0	4.0	5.0	6.0
0.1	0.0047	0.0092	0.0132	0.0168	0.0198	0.0222	0.0242	0.0258	0.0270	0.0279	0.0293	0.0301	0.0306	0.0309	0.0311	0.0314	0.0315	0.0316	0.0316	0.0316
0.2	0.0092	0.0179	0.0259	0.0328	0.0387	0.0435	0.0474	0.0504	0.0528	0.0547	0.0573	0.0589	0.0599	0.0606	0.0610	0.0616	0.0618	0.0619	0.0620	0.0620
0.3	0.0132	0.0259	0.0374	0.0474	0.0559	0.0629	0.0686	0.0731	0.0766	0.0794	0.0832	0.0856	0.0871	0.0880	0.0887	0.0895	0.0898	0.0901	0.0901	0.0902
0.4	0.0168	0.0328	0.0474	0.0602	0.0711	0.0801	0.0873	0.0931	0.0977	0.1013	0.1063	0.1094	0.1114	0.1126	0.1134	0.1145	0.1150	0.1153	0.1154	0.1154
0.5	0.0198	0.0387	0.0559	0.0711	0.0840	0.0947	0.1034	0.1104	0.1158	0.1202	0.1263	0.1300	0.1324	0.1340	0.1350	0.1363	0.1368	0.1372	0.1374	0.1374
0.6	0.0222	0.0435	0.0629	0.0801	0.0947	0.1069	0.1168	0.1247	0.1311	0.1361	0.1431	0.1475	0.1503	0.1521	0.1533	0.1548	0.1555	0.1560	0.1561	0.1562
0.7	0.0242	0.0474	0.0686	0.0873	0.1034	0.1169	0.1277	0.1365	0.1436	0.1491	0.1570	0.1620	0.1652	0.1672	0.1686	0.1704	0.1711	0.1717	0.1719	0.1719
0.8	0.0258	0.0504	0.0731	0.0931	0.1104	0.1247	0.1365	0.1461	0.1537	0.1598	0.1684	0.1739	0.1774	0.1797	0.1812	0.1832	0.1841	0.1847	0.1849	0.1850
0.9	0.0270	0.0528	0.0766	0.0977	0.1158	0.1311	0.1436	0.1537	0.1619	0.1684	0.1777	0.1836	0.1874	0.1899	0.1915	0.1938	0.1947	0.1954	0.1956	0.1957
1.0	0.0279	0.0547	0.0794	0.1013	0.1202	0.1361	0.1491	0.1598	0.1684	0.1752	0.1851	0.1914	0.1955	0.1981	0.1999	0.2024	0.2034	0.2042	0.2044	0.2045
1.2	0.0293	0.0573	0.0832	0.1063	0.1263	0.1431	0.1570	0.1684	0.1777	0.1851	0.1958	0.2028	0.2073	0.2103	0.2124	0.2151	0.2163	0.2172	0.2175	0.2176
1.4	0.0301	0.0589	0.0856	0.1094	0.1300	0.1475	0.1620	0.1739	0.1836	0.1914	0.2028	0.2102	0.2151	0.2184	0.2206	0.2236	0.2250	0.2260	0.2263	0.2264
1.6	0.0306	0.0599	0.0871	0.1114	0.1324	0.1503	0.1652	0.1774	0.1874	0.1955	0.2073	0.2151	0.2203	0.2237	0.2261	0.2294	0.2309	0.2320	0.2323	0.2325
1.8	0.0309	0.0606	0.0880	0.1126	0.1340	0.1521	0.1672	0.1797	0.1899	0.1981	0.2103	0.2183	0.2237	0.2274	0.2299	0.2333	0.2350	0.2362	0.2366	0.2367
2.0	0.0311	0.0610	0.0887	0.1134	0.1350	0.1533	0.1686	0.1812	0.1915	0.1999	0.2124	0.2206	0.2261	0.2299	0.2325	0.2361	0.2378	0.2391	0.2395	0.2397
2.5	0.0314	0.0616	0.0895	0.1145	0.1363	0.1548	0.1704	0.1832	0.1938	0.2024	0.2151	0.2236	0.2294	0.2333	0.2361	0.2401	0.2420	0.2434	0.2439	0.2441
3.0	0.0315	0.0618	0.0898	0.1150	0.1368	0.1555	0.1711	0.1841	0.1947	0.2034	0.2163	0.2250	0.2309	0.2350	0.2378	0.2420	0.2439	0.2455	0.2460	0.2463
4.0	0.0316	0.0619	0.0901	0.1153	0.1372	0.1560	0.1717	0.1847	0.1954	0.2042	0.2172	0.2260	0.2320	0.2362	0.2391	0.2434	0.2455	0.2472	0.2479	0.2481
5.0	0.0316	0.0620	0.0901	0.1154	0.1374	0.1561	0.1719	0.1849	0.1956	0.2044	0.2175	0.2263	0.2324	0.2366	0.2395	0.2439	0.2460	0.2479	0.2486	0.2489
6.0	0.0316	0.0620	0.0902	0.1154	0.1374	0.1562	0.1719	0.1850	0.1957	0.2045	0.2176	0.2264	0.2325	0.2367	0.2397	0.2441	0.2463	0.2482	0.2489	0.2492

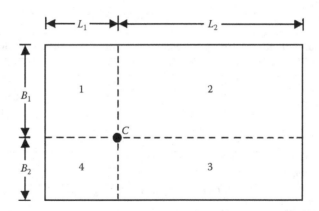

FIGURE 5.7 Stress increase below any point of a uniformly loaded flexible rectangular area.

and 4, $m_2 = B_1/z$; $n_2 = L_2/z$, $m_3 = B_2/z$; $n_3 = L_2/z$, and $m_4 = B_2/z$; $n_4 = L_1/z$. Now, using Table 5.2, the magnitudes of $I (= I_1, I_2, I_3, I_4)$ for the four rectangles can be determined. The total stress increase below point C at depth z can thus be determined as

$$\sigma_z = q(I_1 + I_2 + I_3 + I_4) \tag{5.16}$$

In most practical problems the stress increase below the center of a loaded rectangular area is of primary importance. The vertical stress increase below the *center* of a uniformly loaded *flexible* rectangular area can be calculated as

$$\sigma_{z(c)} = \frac{2q}{\pi}\left[\frac{m_1 n_1}{\sqrt{1+m_1^2+n_1^2}}\frac{1+m_1^2+n_1^2}{(1+n_1^2)(m_1^2+n_1^2)} + \sin^{-1}\frac{m_1}{\sqrt{m_1^2+n_1^2}\sqrt{1+n_1^2}}\right] \tag{5.17}$$

where

$$m_1 = \frac{L}{B} \tag{5.18}$$

$$n_1 = \frac{z}{\left(\frac{B}{2}\right)} \tag{5.19}$$

Table 5.3 gives the variation of $\sigma_{z(c)}/q$ with L/B and z/B based on equation (5.17).

EXAMPLE 5.1

Figure 5.8 shows the plan of a flexible loaded area located at the ground surface. The uniformly distributed load q on the area is 150 kN/m². Determine the stress increase σ_z below points A and C at a depth of 10 m below the ground surface. Note that C is at the center of the area.

Solution

Stress increase below point A.
The following table can now be prepared:

Area No.	$B(m)$	$L(m)$	$z(m)$	$m = B/z$	$n = B/z$	I (Table 5.2)
1	2	2	10	0.2	0.2	0.0179
2	2	4	10	0.2	0.4	0.0328
3	2	4	10	0.2	0.4	0.0328
4	2	2	10	0.2	0.2	0.0179

Note: $\Sigma 0.1014 = \Sigma T$

From equation (5.14),

$$\sigma_z = qI = (150)(0.1014) = \mathbf{15.21 \ kN/m^2}$$

Stress increase below point C:

$$\frac{L}{B} = \frac{6}{4} = 1.5; \quad \frac{z}{B} = \frac{10}{4} = 2.5$$

From Table 5.3,

$$\frac{\sigma_z}{q} \approx 0.104$$

$$\sigma_z = (0.104)(150) = \mathbf{15.6 \ kN/m^2}$$

TABLE 5.3
Variation of $\sigma_{z\,(c)}/q$ [Equation 5.17)]

	L/B									
z/B	1	2	3	4	5	6	7	8	9	10
0.1	0.994	0.997	0.997	0.997	0.997	0.997	0.997	0.997	0.997	0.997
0.2	0.960	0.976	0.977	0.977	0.977	0.977	0.977	0.977	0.977	0.977
0.3	0.892	0.932	0.936	0.936	0.937	0.937	0.937	0.937	0.937	0.937
0.4	0.800	0.870	0.878	0.880	0.881	0.881	0.881	0.881	0.881	0.881
0.5	0.701	0.800	0.814	0.817	0.818	0.818	0.818	0.818	0.818	0.818
0.6	0.606	0.727	0.748	0.753	0.754	0.755	0.755	0.755	0.755	0.755
0.7	0.522	0.658	0.685	0.692	0.694	0.695	0.695	0.696	0.696	0.696
0.8	0.449	0.593	0.627	0.636	0.639	0.640	0.641	0.641	0.641	0.642
0.9	0.388	0.534	0.573	0.585	0.590	0.591	0.592	0.592	0.593	0.593
1.0	0.336	0.481	0.525	0.540	0.545	0.547	0.548	0.549	0.549	0.549
1.5	0.179	0.293	0.348	0.373	0.384	0.389	0.392	0.393	0.394	0.395
2.0	0.108	0.190	0.241	0.269	0.285	0.293	0.298	0.301	0.302	0.303
2.5	0.072	0.131	0.174	0.202	0.219	0.229	0.236	0.240	0.242	0.244
3.0	0.051	0.095	0.130	0.155	0.172	0.184	0.192	0.197	0.200	0.202
3.5	0.038	0.072	0.100	0.122	0.139	0.150	0.158	0.164	0.168	0.171
4.0	0.029	0.056	0.079	0.098	0.113	0.125	0.133	0.139	0.144	0.147
4.5	0.023	0.045	0.064	0.081	0.094	0.105	0.113	0.119	0.124	0.128
5.0	0.019	0.037	0.053	0.067	0.079	0.089	0.097	0.103	0.108	0.112

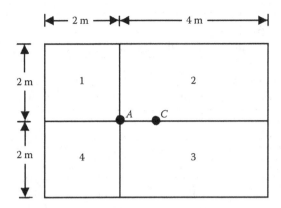

FIGURE 5.8 Uniformly loaded flexible rectangular area.

5.3　STRESS INCREASE DUE TO APPLIED LOAD—WESTERGAARD'S SOLUTION

5.3.1　Point Load

Westergaard[3] proposed a solution for determining the vertical stress caused by a point load Q in an elastic solid medium in which layers alternate with thin rigid reinforcements. This type of assumption may be an idealization of a clay layer with thin seams of sand. For such an assumption, the vertical stress increase at a point A (Figure 5.2) can be given by

$$\sigma_z = \frac{Q\eta}{2\pi z^2}\left[\frac{1}{\eta^2 + \left(\frac{r}{z}\right)^2}\right]^{-3/2} \tag{5.20}$$

where

$$\eta = \sqrt{\frac{1-2v}{2-2v}} \tag{5.21}$$

v = Poisson's ratio of the solid between the rigid reinforcements
$r = \sqrt{x^2 + y^2}$

Equation (5.20) can be rewritten as

$$\sigma_z = \frac{Q}{z^2}I' \tag{5.22}$$

where

$$I' = \frac{1}{2\pi\eta^2}\left[\left(\frac{r}{\eta z}\right)^2 + 1\right]^{-3/2} \tag{5.23}$$

The variations of I' with r/z and v are given in Table 5.4.

TABLE 5.4

Variation of I' with r/z and v [Equation (5.23)]

r/z	I'		
	$v = 0$	$v = 0.2$	$v = 0.4$
0	0.3183	0.4244	0.9550
0.25	0.2668	0.3368	0.5923
0.50	0.1733	0.1973	0.2416
0.75	0.1028	0.1074	0.1044
1.00	0.0613	0.0605	0.0516
1.25	0.0380	0.0361	0.0286
1.50	0.0247	0.0229	0.0173
1.75	0.0167	0.0153	0.0112
2.00	0.0118	0.0107	0.0076
2.25	0.0086	0.0077	0.0054
2.50	0.0064	0.0057	0.0040
2.75	0.0049	0.0044	0.0030
3.00	0.0038	0.0034	0.0023
3.25	0.0031	0.0027	0.0019
3.50	0.0025	0.0022	0.0015
3.75	0.0021	0.0018	0.0012
4.00	0.0017	0.0015	0.0010
4.25	0.0014	0.0012	0.0008
4.50	0.0012	0.0010	0.0007
4.75	0.0010	0.0009	0.0006
5.00	0.0009	0.0008	0.0005

5.3.2 UNIFORMLY LOADED FLEXIBLE CIRCULAR AREA

Refer to Figure 5.4, which shows a uniformly loaded flexible circular area of radius R. If the circular area is located on a Westergaard-type material, the increase in vertical stress σ_z at a point located at a depth z immediately below the center of the area can be given as

$$\sigma_z = q \left\{ 1 - \frac{\eta}{\left[\eta^2 + \left(\frac{R}{z} \right)^2 \right]^{1/2}} \right\}$$ (5.24)

The variations of σ_z/q with R/z and $v = 0$ are given in Table 5.5.

5.3.3 UNIFORMLY LOADED FLEXIBLE RECTANGULAR AREA

Refer to Figure 5.6. If the flexible rectangular area is located on a Westergaard-type material, the stress increase at a point A can be given as

$$\sigma_z = \frac{q}{2\pi} \left[\cot^{-1} \sqrt{\eta^2 \left(\frac{1}{m^2} + \frac{1}{n^2} \right) + \eta^4 \left(\frac{1}{m^2 n^2} \right)} \right]$$ (5.25)

TABLE 5.5
Variation of σ_z/q with R/z
and $v = 0$ [Equation (5.24)]

R/z	σ_z/q
0	0
0.25	0.0572
0.50	0.1835
0.75	0.3140
1.00	0.4227
1.25	0.5076
1.50	0.5736
1.75	0.6254
2.00	0.6667
2.25	0.7002
2.50	0.7278
2.75	0.7510
3.00	0.7706
4.00	0.8259
5.00	0.8600

where

$$m = \frac{B}{z}$$

$$n = \frac{L}{z}$$

5.4 ELASTIC SETTLEMENT

5.4.1 FLEXIBLE AND RIGID FOUNDATIONS

Before discussing the relationships for elastic settlement of shallow foundations, it is important to understand the fundamental concepts and the differences between a flexible foundation and a rigid foundation. When a flexible foundation on an *elastic medium* is subjected to a uniformly distributed load, the contact pressure will be uniform, as shown in Figure 5.9a. Figure 5.9a also shows the settlement profile of the foundation. If a similar foundation is placed on granular soil it will undergo larger elastic settlement at the edges rather than at the center (Figure 5.9b); however, the contact pressure will be uniform. The larger settlement at the edges is due to the lack of confinement in the soil.

If a fully rigid foundation is placed on the surface of elastic medium, the settlement will remain the same at all points; however, the contact distribution will be as shown in Figure 5.10a. If this rigid foundation is placed on granular soil, the contact pressure distribution will be as shown in Figure 5.10b, although the settlement at all points below the foundation will be the same.

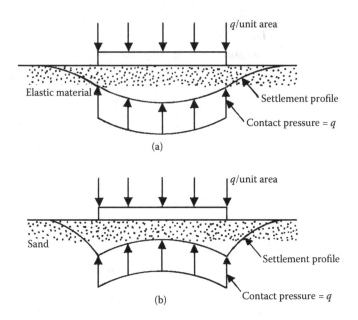

FIGURE 5.9 Contact pressures and settlements for a flexible foundation: (a) elastic material; (b) granular soil.

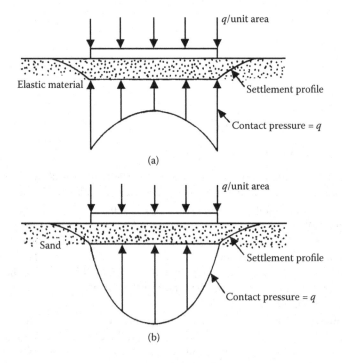

FIGURE 5.10 Contact pressures and settlements for a rigid foundation: (a) elastic material; (b) granular soil.

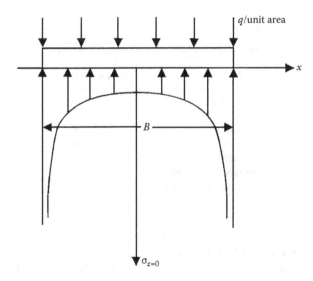

FIGURE 5.11 Contact pressure distributions under an infinitely rigid foundation supported by a perfectly elastic material.

Theoretically, for an *infinitely rigid* foundation supported by a *perfectly elastic material*, the contact pressure can be expressed as (Figure 5.11)

$$\sigma_{z=0} = \frac{2q}{\pi\sqrt{1-\left(\dfrac{2x}{B}\right)^2}} \quad \text{(continuous foundation)} \tag{5.26}$$

$$\sigma_{z=0} = \frac{q}{2\sqrt{1-\left(\dfrac{2x}{B}\right)^2}} \quad \text{(circular foundation)} \tag{5.27}$$

where

q = applied load per unit area of the foundation
B = foundation width (or diameter)

Borowicka[4] developed solutions for the distribution of contact pressure beneath a continuous foundation supported by a perfectly elastic material. According to his theory,

$$\sigma_{z=0} = f(K) \tag{5.28}$$

where

$$K = \text{relative stiffness factor} = \frac{1}{6}\left(\frac{1-v_s^2}{1-v_f^2}\right)\left(\frac{E_f}{E_s}\right)\left(\frac{t}{\dfrac{B}{2}}\right)^3 \tag{5.29}$$

v_s = Poisson's ratio of the elastic material
v_f = Poisson's ratio of the foundation material

TABLE 5.6
Suggested Values for Poisson's Ratio

Soil Type	Poisson's Ratio v
Coarse sand	0.15–0.20
Medium loose sand	0.20–0.25
Fine sand	0.25–0.30
Sandy silt and silt	0.30–0.35
Saturated clay (undrained)	0.50
Saturated clay—lightly overconsolidated (drained)	0.2–0.4

t = thickness of the foundation

E_s, E_f = modulus of elasticity of the elastic material and foundation material, respectively

Although soil is not perfectly elastic and homogeneous, the theory of elasticity may be used to estimate the settlements of shallow foundations at allowable loads. Judicious uses of these results have done well in the design, construction, and maintenance of structures.

5.4.2 ELASTIC PARAMETERS

Parameters such as the modulus of elasticity E_s and Poisson's ratio v for a given soil must be known in order to calculate the elastic settlement of a foundation. In most cases, if laboratory test results are not available, they are estimated from empirical correlations. Table 5.6 provides some suggested values for Poisson's ratio.

Trautmann and Kulhawy[5] used the following relationship for Poisson's ratio (drained state):

$$v = 0.1 + 0.3\phi_{rel} \tag{5.30}$$

$$\phi_{rel} = \text{relative friction angle} = \frac{\phi_{tc} - 25°}{45° - 25°} \quad (0 \le \phi_{rel} \le 1) \tag{5.31}$$

where

ϕ_{tc} = friction angle from drained triaxial compression test

A general range of the modulus of elasticity of sand E_s is given in Table 5.7.

A number of correlations for the modulus of elasticity of sand with the field standard penetration resistance N_{60} and cone penetration resistance q_c have been made in the past. Schmertmann[6] proposed that

$$E_s \ (kN/m^2) = 766N_{60} \tag{5.32}$$

TABLE 5.7

General Range of Modulus of Elasticity of Sand

Type	E_s (kN/m²)
Coarse and medium coarse sand	
Loose	25,000–35,000
Medium dense	30,000–40,000
Dense	40,000–45,000
Fine sand	
Loose	20,000–25,000
Medium dense	25,000–35,000
Dense	35,000–40,000
Sandy silt	
Loose	8,000–12,000
Medium dense	10,000–12,000
Dense	12,000–15,000

Schmertmann et al.[7] made the following recommendations for estimating the E_s of sand from cone penetration resistance, or

$$E_s = 2.5q_c \quad \text{(for square and circular foundations)} \quad (5.33)$$

$$E_s = 3.5q_c \quad \text{(for strip foundations; } L/B \geq 10) \quad (5.34)$$

In many cases, the modulus of elasticity of *saturated clay* soils (undrained) has been correlated with the undrained shear strength c_u. D'Appolonia et al.[8] compiled several field test results and concluded that

$$\frac{E_s}{c_u} = 1000 \text{ to } 1500 \quad \left(\begin{array}{c} \text{for lean inorganic clays from} \\ \text{moderate to high plasticity} \end{array} \right) \quad (5.35)$$

Duncan and Buchignani[9] correlated E_s/c_u with the overconsolidation ratio OCR and plasticity index *PI* of several clay soils. This broadly generalized correlation is shown in Figure 5.12.

5.4.3 SETTLEMENT OF FOUNDATIONS ON SATURATED CLAYS

Janbu et al.[10] proposed a generalized equation for estimating the average elastic settlement of a uniformly loaded flexible foundation located on saturated clay ($v = 0.5$). This relationship incorporates (a) the effect of embedment D_f, and (b) the possible existence of a rigid layer at a shallow depth under the foundation as shown in Figure 5.13, or,

$$S_e = \mu_1 \mu_2 \frac{qB}{E_s} \quad (5.36)$$

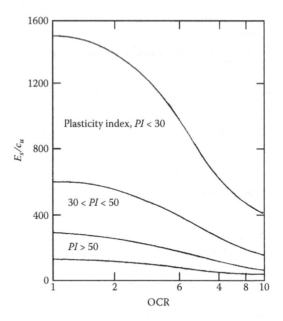

FIGURE 5.12 Correlation of Duncan and Buchignani for the modulus of elasticity of clay in an undrained state.

where

$$\mu_1 = f\left(\frac{D_f}{B}\right)$$

$$\mu_2 = \left(\frac{H}{B}, \frac{L}{B}\right)$$

L = foundation length
B = foundation width

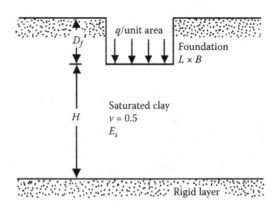

FIGURE 5.13 Settlement of foundation on saturated clay.

TABLE 5.8
Variation of μ_1 with D_f/B
[Equation (5.36)]

D_f/B	μ_1
0	1.0
2	0.9
4	0.88
6	0.875
8	0.87
10	0.865
12	0.863
14	0.860
16	0.856
18	0.854
20	0.850

Christian and Carrier[11] made a critical evaluation of the factors μ_1 and μ_2, and the results were presented in graphical form. The interpolated values of μ_1 and μ_2 from these graphs are given in Tables 5.8 and 5.9.

5.4.4 FOUNDATIONS ON SAND—CORRELATION WITH STANDARD PENETRATION RESISTANCE

There are several empirical relationships to estimate the elastic settlements of foundations on granular soil that are based on the correlations with the width of the foundation and the standard penetration resistance obtained from the field, N_{60} (that is, penetration resistance with an average energy ratio of 60%). Some of these correlations are outlined in this section.

TABLE 5.9
Variation of μ_2 with H/B and L/B [Equation (5.36)]

H/B			L/B			
	Circle	1	2	5	10	∞
1	0.36	0.36	0.36	0.36	0.36	0.36
2	0.47	0.53	0.63	0.64	0.64	0.64
4	0.58	0.63	0.82	0.94	0.94	0.94
6	0.61	0.67	0.88	1.08	1.14	1.16
8	0.62	0.68	0.90	1.13	1.22	1.26
10	0.63	0.70	0.92	1.18	1.30	1.42
20	0.64	0.71	0.93	1.26	1.47	1.74
30	0.66	0.73	0.95	1.29	1.54	1.84

5.4.4.1 Terzaghi and Peck's Correlation

Terzaghi and Peck[12] proposed the following empirical relationship between the settlement S_e of a prototype foundation measuring $B \times B$ in plan and the settlement of a test plate $S_{e(1)}$ measuring $B_1 \times B_1$ loaded to the same intensity:

$$\frac{S_e}{S_{e(1)}} = \frac{4}{\left[1 + \left(\dfrac{B_1}{B}\right)^2\right]^2} \tag{5.37}$$

Although a full-sized footing can be used for a load test, the normal practice is to employ a plate of the order of $B_1 = 0.3$ m to 1 m. Terzaghi and Peck[12] also proposed a correlation for the allowable bearing capacity, standard penetration number N_{60}, and the width of the foundation B corresponding to a 25-mm settlement based on the observations given by equation (5.37). The curves that give the preceding correlation can be approximated by the relation,

$$S_e \text{ (mm)} = \frac{3q}{N_{60}} \left(\frac{B}{B + 0.3}\right)^2 \tag{5.38}$$

where
q = bearing pressure in kN/m^2
B = width of foundation in m

If corrections for groundwater table location and depth of embedment are included, then equation (5.38) takes the form,

$$S_e = C_W C_D \frac{3q}{N_{60}} \left(\frac{B}{B + 0.3}\right)^2 \tag{5.39}$$

where
C_W = groundwater table correction
C_D = correction for depth of embedment $= 1 - \left(\dfrac{D_f}{4B}\right)$
D_f = depth of embedment

The magnitude of C_W is equal to 1.0 if the depth of the water table is greater than or equal to $2B$ below the foundation, and it is equal to 2.0 if the depth of the water table is less than or equal to B below the foundation. The N_{60} values used in equations (5.38) and (5.39) should be the average value of N_{60} up to a depth of about $3B$ to $4B$ measured from the bottom of the foundation.

5.4.4.2 Meyerhof's Correlation

In 1956, Meyerhof[13] proposed the following relationships for S_e:

$$S_e = \frac{2q}{N_{60}} \quad \text{(for } B \leq 1.22 \text{ m)} \tag{5.40a}$$

and

$$S_e = \frac{3q}{N_{60}} \left(\frac{B}{B+0.3} \right)^2 \quad \text{(for } B > 1.22 \text{ m)} \tag{5.40b}$$

where S_e is in mm, B is in m, and q is in kN/m^2.

Note that equations (5.38) and (5.40b) are similar. In 1965, Meyerhof[14] compared the predicted and observed settlements of eight structures and proposed revisions to equations (5.40a) and (5.40b). According to these revisions,

$$S_e \approx \frac{1.25q}{N_{60}} \quad \text{(for } B \leq 1.22 \text{ m)} \tag{5.41}$$

and

$$S_e = \frac{2q}{N_{60}} \left(\frac{B}{B+0.3} \right)^2 \quad \text{(for } B > 1.22 \text{ m)} \tag{5.42}$$

Comparing equations (5.40a) and (5.40b) with equations (5.41) and (5.42) it can be seen that, for similar settlement levels, the allowable pressure q is 50% higher for equations (5.41) and (5.42). If corrections for the location of the groundwater table and depth of embedment are incorporated into equations (5.41) and (5.42), we obtain

$$S_e \text{ (mm)} = C_W C_D \frac{1.25q}{N_{60}} \quad \text{(for } B \leq 1.22 \text{ m)} \tag{5.43}$$

and

$$S_e \text{ (mm)} = C_W C_D \frac{2q}{N_{60}} \left(\frac{B}{B+0.3} \right)^2 \quad \text{(for } B > 1.22 \text{ m)} \tag{5.44}$$

$$C_W = 1.0 \tag{5.45}$$

and

$$C_D = 1.0 - \frac{D_f}{4B} \tag{5.46}$$

5.4.4.3 Peck and Bazaraa's Method

The original work of Terzaghi and Peck[12] as given in equation (5.38) was subsequently compared to several field observations. It was found that the relationship provided by equation (5.38) is overly conservative (that is, observed field settlements were substantially lower than those predicted by the equation). Recognizing this fact, Peck and Bazaraa[15] suggested the following revision to equation (5.39):

$$S_e = C_W C_D \frac{2q}{(N_1)_{60}} \left(\frac{B}{B+0.3} \right)^2 \tag{5.47}$$

where

S_e is in mm, q is in kN/m², and B is in m

$(N_1)_{60}$ = corrected standard penetration number

$$C_W = \frac{\sigma_o \text{ at } 0.5B \text{ below the bottom of the foundation}}{\sigma'_o \text{ at } 0.5B \text{ below the bottom of the foundation}} \tag{5.48}$$

σ_o = total overburden pressure

σ'_o = effective overburden pressure

$$C_D = 1.0 - 0.4\left(\frac{\gamma D_f}{q}\right)^{0.5} \tag{5.49}$$

γ = unit weight of soil

The relationships for $(N_1)_{60}$ are as follows:

$$(N_1)_{60} = \frac{4N_{60}}{1 + 0.04\sigma'_o} \quad \text{(for } \sigma'_o \leq 75 \text{ kN/m}^2\text{)} \tag{5.50}$$

and

$$(N_1)_{60} = \frac{4N_{60}}{3.25 + 0.01\sigma'_o} \quad \text{(for } \sigma'_o > 75 \text{ kN/m}^2\text{)} \tag{5.51}$$

where

σ'_o = the effective overburden pressure

5.4.4.4 Burland and Burbidge's Method

Burland and Burbidge[16] proposed a method for calculating the elastic settlement of sandy soil using the field standard penetration number N_{60}. According to this procedure, following are the steps to estimate the elastic settlement of a foundation:

1. Determination of Variation of Standard Penetration Number with Depth
   The Obtain the field penetration numbers N_{60} with depth at the location of the foundation. Depending on the field conditions, the following adjustments of N_{60} may be necessary:
   For gravel or sandy gravel,

$$N_{60(a)} \approx 1.25 N_{60} \tag{5.52}$$

   For fine sand or silty sand below the groundwater table and $N_{60} > 15$,

$$N_{60(a)} \approx 15 + 0.5(N_{60} - 15) \tag{5.53}$$

where

$N_{60(a)}$ = adjusted N_{60} value

2. Determination of Depth of Stress Influence z'

In determining the depth of stress influence, the following three cases may arise:

Case I. If N_{60} [or $N_{60(a)}$] is approximately constant with depth, calculate z' from

$$\frac{z'}{B_R} = 1.4\left(\frac{B}{B_R}\right)^{0.75} \tag{5.54}$$

where

B_R = reference width = 0.3 m

B = width of the actual foundation (m)

Case II. If N_{60} [or $N_{60(a)}$] is increasing with depth, use equation (5.54) to calculate z'.

Case III. If N_{60} [or $N_{60(a)}$] is decreasing with depth, calculate $z' = 2B$ and $z' =$ distance from the bottom of the foundation to the bottom of the soft soil layer ($= z''$). Use $z' = 2B$ or $z' = z''$ (whichever is smaller).

3. Determination of Depth of Influence Correction Factor α

The correction factor α is given as (Note: H = depth of comparable soil layer)

$$\alpha = \frac{H}{z'}\left(2 - \frac{H}{z'}\right) \le 1 \tag{5.55}$$

4. Calculation of Elastic Settlement

The elastic settlement of the foundation S_e can be calculated as

$$\frac{S_e}{B_R} = 0.14\alpha\left\{\frac{1.71}{[\bar{N}_{60} \text{ or } \bar{N}_{60(a)}]^{1.4}}\right\}\left[\frac{1.25\left(\frac{L}{B}\right)}{0.25 + \left(\frac{L}{B}\right)}\right]^2\left(\frac{B}{B_R}\right)^{0.7}\left(\frac{q}{p_a}\right)$$

(for normally consolidated soil) $\tag{5.56}$

where

L = length of the foundation

p_a = atmospheric pressure ($\approx$ 100 kN/m²)

$\bar{N}_{60}$ or $\bar{N}_{60(a)}$ = average value of N_{60} or $N_{60(a)}$ in the depth of stress influence

$$\frac{S_e}{B_R} = 0.047\alpha\left\{\frac{0.57}{[\bar{N}_{60} \text{ or } \bar{N}_{60(a)}]^{1.4}}\right\}\left[\frac{1.25\left(\frac{L}{B}\right)}{0.25 + \left(\frac{L}{B}\right)}\right]^2\left(\frac{B}{B_R}\right)^{0.7}\left(\frac{q}{p_a}\right)$$

For overconsolidated soil ($q \le \sigma'_c$ where σ'_c = overconsolidation pressure)

$\tag{5.57}$

$$\frac{S_e}{B_R} = 0.14\alpha\left\{\frac{0.57}{[\bar{N}_{60} \text{ or } \bar{N}_{60(a)}]^{1.4}}\right\}\left[\frac{1.25\left(\frac{L}{B}\right)}{0.25 + \left(\frac{L}{B}\right)}\right]^2\left(\frac{B}{B_R}\right)^{0.7}\left(\frac{q - 0.67\sigma'_c}{p_a}\right)$$

For overconsolidated soil ($q > \sigma'_c$) $\tag{5.58}$

EXAMPLE 5.2

A shallow foundation measuring $1.75 \text{ m} \times 1.75 \text{ m}$ is to be constructed over a layer of sand. Given: $D_f = 1$ m; N_{60} is generally increasing with depth; $\bar{N}_{60}$ in the depth of stress influence = 10; $q = 120$ kN/m². The sand is normally consolidated. Estimate the elastic settlement of the foundation. Use the Burland and Burbidge method.

Solution

From equation (5.54),

$$\frac{z'}{B_R} = 1.4 \left(\frac{B}{B_R} \right)^{0.75}$$

the depth of stress influence is

$$z' = 1.4 \left(\frac{B}{B_R} \right)^{0.75} B_R = (1.4) \left(\frac{1.75}{0.3} \right)^{0.75} (0.3) \approx 1.58 \text{ m}$$

From equation (5.55), $\alpha = 1$. From equation (5.56) (note $L/B = 1$; $p_a \approx 100$ kN/m²),

$$\frac{S_e}{B_R} = 0.14 \alpha \left\{ \frac{1.71}{(\bar{N}_{60})^{1.4}} \right\} \left[\frac{1.25 \left(\frac{L}{B} \right)}{0.25 + \left(\frac{L}{B} \right)} \right]^2 \left(\frac{B}{B_R} \right)^{0.7} \left(\frac{q}{p_a} \right)$$

$$= (0.14)(1) \left\{ \frac{1.71}{(10)^{1.4}} \right\} \left[\frac{1.25(1)}{0.25 + (1)} \right]^2 \left(\frac{1.75}{0.3} \right)^{0.7} \left(\frac{120}{100} \right) = 0.0118 \text{ m} = \mathbf{11.8 \text{ mm}}$$

EXAMPLE 5.3

Solve the problem in Example 5.2 using Meyerhof's method.

Solution

From equation (5.44),

$$S_e = C_W C_D \frac{2q}{N_{60}} \left(\frac{B}{B + 0.3} \right)^2$$

$$C_W = 1$$

$$C_D = 1 - \left(\frac{D_f}{4B} \right) = 1 - \frac{1}{(4)(1.75)} \approx 0.86$$

$$S_e = (0.86)(1) \frac{(2)(120)}{10} \left(\frac{1.75}{1.75 + 0.3} \right)^2 = \mathbf{15.04 \text{ mm}}$$

5.4.5 FOUNDATIONS ON GRANULAR SOIL—USE OF STRAIN INFLUENCE FACTOR

Referring to Figure 5.4, the equation for vertical strain ε_z below the center of a flexible circular load of radius R can be given as

$$\varepsilon_z = \frac{1}{E_s}[\sigma_Z - v(\sigma_r + \sigma_\theta)] \qquad (5.59)$$

After proper substitution for σ_z, σ_r, and σ_θ in the preceding equation, one obtains

$$\varepsilon_z = \frac{q(1+v)}{E_s}[(1-2v)A' + B'] \qquad (5.60)$$

where
A', B' = nondimensional factors and functions of z/R

The variations of A' and B' below the center of a loaded area as estimated by Ahlvin and Ulery[2] are given in Table 5.10. From equation (5.60) we can write

$$I_z = \frac{\varepsilon_z E_s}{q}(1+v)[(1-2v)A' + B'] \qquad (5.61)$$

Figure 5.14 shows plots of I_z versus z/R obtained from the experimental results of Eggestad[17] along with the theoretical values calculated from equation (5.61). Based on Figure 5.14, Schmertmann[6] proposed a practical variation of I_z and z/B

TABLE 5.10

Variations of A' and B'
(Below the Center of a
Flexible Loaded Area)

z/R	A'	B'
0	1.0	0
0.2	0.804	0.189
0.4	0.629	0.320
0.6	0.486	0.378
0.8	0.375	0.381
1.0	0.293	0.354
1.5	0.168	0.256
2.0	0.106	0.179
2.5	0.072	0.128
3.0	0.051	0.095
4.0	0.030	0.057
5.0	0.019	0.038
6.0	0.014	0.027
7.0	0.010	0.020
8.0	0.008	0.015
9.0	0.006	0.012

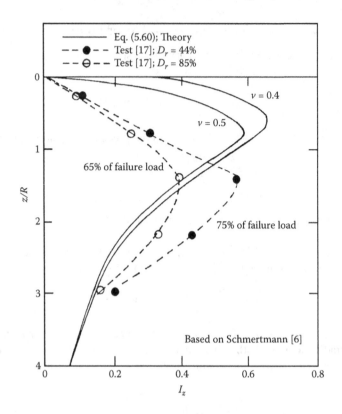

FIGURE 5.14 Comparison of experiment and theoretical variations of I_z below the center of a flexible circularly loaded area. *Note:* R = radius of circular area; D_r = relative density.

(B = foundation width) for calculating the elastic settlements of foundations. This model was later modified by Schmertmann et al.,[7] and the variation is shown in Figure 5.15 for $L/B = 1$ and $L/B \geq 10$. Interpolations can be used to obtain the $I_z - z/B$ variations for other L/B values. Using the simplified strain influence factor, the elastic settlement can be calculated as

$$S_e = c_1 c_2 (q - q') \sum \left(\frac{I_z}{E_s} \right) \Delta z \qquad (5.62)$$

where

c_1 = a correction factor for depth of foundation $= 1 - 0.5 \left(\dfrac{q'}{q - q'} \right)$

c_2 = a correction factor for creep in soil $= 1 + 0.2 \log \left(\dfrac{\text{time in years}}{0.1} \right)$

$q' = \gamma D_f$
q = stress at the level of the foundation

The use of equation (5.62) can be explained by the following example.

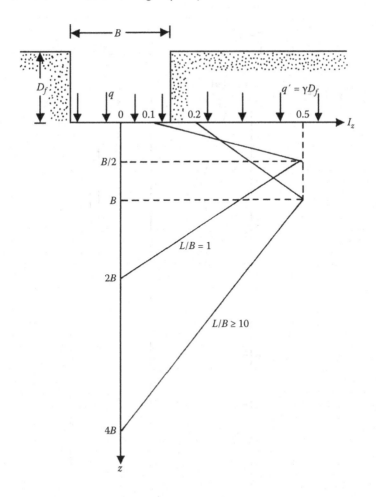

FIGURE 5.15 Variation of I_z versus z/B.

EXAMPLE 5.4

Figure 5.16a shows a continuous foundation for which $B = 2$ m; $D_f = 1$ m; unit weight of sand $\gamma = 17$ kN/m³; $q = 175$ kN/m². For this case, L/B is greater than 10. Accordingly, the plot of I_z with depth is shown in Figure 5.16a. Note that: $I_z = 0.2$ at $z = 0$; $I_z = 0.5$ at $z = 2$ m ($= B$), and $I_z = 0$ at $z = 8$ m ($= 4B$). Based on the results of the standard penetration test or cone penetration test, the variation of E_s can be calculated using equation (5.32) or (5.34) (or similar relationships). The variation is shown by the dashed line in Figure 5.16b. The actual variation of E_s can be approximated by several linear plots, and this is also shown in Figure 5.16b (solid lines). For elastic settlement, Table 5.11 can now be prepared. Since $\gamma = 17$ kN/m³, $q' = \gamma D_f = (1)(17) = 17$ kN/m². Given: $q = 175$ kN/m². Thus, $q - q' = 175 - 17 = 158$ kN/m². Also,

$$c_1 = 1 - 0.5\left(\frac{q'}{q - q'}\right) = 1 - 0.5\left(\frac{17}{158}\right) = 0.946$$

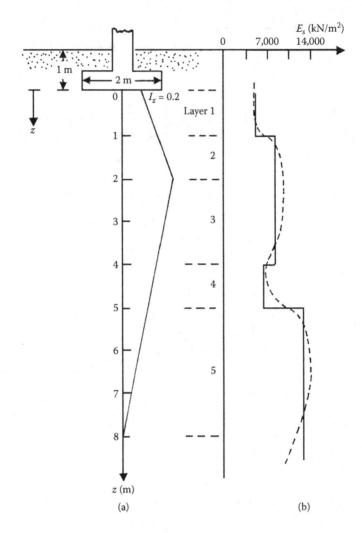

FIGURE 5.16 Determination of elastic settlement of a continuous foundation by strain influence factor method.

Assume the time for creep is 10 years. Hence,

$$c_2 = 1 + 0.2\log\left(\frac{10}{0.1}\right) = 1.4$$

Thus,

$$S_e = c_1 c_2 (q - q') \sum\left(\frac{I_z}{E_s}\right)\Delta z = (0.946)(1.4)(158)(26.45 \times 10^{-5}) = 5534.8 \times 10^{-5}\ \text{m} \approx \mathbf{55.35\ mm}$$

TABLE 5.11
Elastic Settlement Calculations (Figure 5.16)

Layer No.	Δz (m)	E_s (kN/m²)	z to the Middle of the Layer (m)	I_z at the Middle of the Layer	$\dfrac{I_z}{E_s}\Delta z$ (m³/kN)
1	1	5250	0.5	0.275	5.23×10^{-5}
2	1	8750	1.5	0.425	4.85×10^{-5}
3	2	8750	3.0	0.417	9.53×10^{-5}
4	1	7000	4.5	0.292	4.17×10^{-5}
5	3	14,000	6.5	0.125	2.67×10^{-5}

Note: $\Sigma 8 \text{ m} = 4B$ $\Sigma\ 26.45 \times 10^{-5}$ m³/kN

5.4.6 FOUNDATIONS ON GRANULAR SOIL—SETTLEMENT CALCULATION BASED ON THEORY OF ELASTICITY

Figure 5.17 shows a schematic diagram of the elastic settlement profile for a flexible and rigid foundation. The shallow foundation measures $B \times L$ in plan and is located at a depth D_f below the ground surface. A rock (or a rigid layer) is located at a depth H below the bottom of the foundation. Theoretically, if the foundation is perfectly

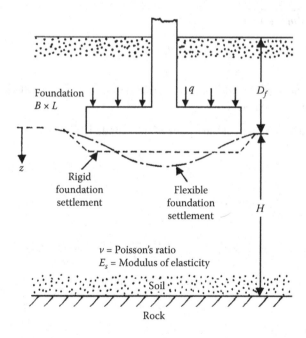

FIGURE 5.17 Settlement profile for shallow flexible and rigid foundations.

flexible (Bowles[18]), the settlement may be expressed as

$$S_e = q(\alpha' B') \frac{1-v^2}{E_s} I_s I_f$$

(5.63)

where

    q = net applied pressure on the foundation

    v = Poisson's ratio of soil

    E_s = average modulus of elasticity of the soil under the foundation measured from $z = 0$ to about $z = 4B$

    $B' = B/2$ for center of foundation ($= B$ for corner of foundation)

$$I_s = \text{shape factor (Steinbrenner}^{[19]}) = F_1 + \frac{1-2v}{1-v} F_2$$

(5.64)

$$F_1 = \frac{1}{\pi}(A_0 + A_1)$$

(5.65)

$$F_2 = \frac{n'}{2\pi} \tan^{-1} A_2$$

(5.66)

$$A_0 = m' \ln \frac{(1+\sqrt{m'^2+1})\sqrt{m'^2+n'^2}}{m'(1+\sqrt{m'^2+n'^2+1})}$$

(5.67)

$$A_1 = \ln \frac{(m'+\sqrt{m'^2+1})\sqrt{1+n'^2}}{m'+\sqrt{m'^2+n'^2+1}}$$

(5.68)

$$A_2 = \frac{m'}{n'+\sqrt{m'^2+n'^2+1}}$$

(5.69)

$$I_f = \text{depth factor (Fox}^{[20]}) = f\left(\frac{D_f}{B}, v, \text{ and } \frac{L}{B}\right)$$

(5.70)

    α' = a factor that depends on the location on the foundation where settlement is being calculated

To calculate settlement at the center of the foundation, we use

$$\alpha' = 4$$

(5.71)

$$m' = \frac{L}{B}$$

(5.72)

and

$$n' = \frac{H}{\left(\frac{B}{2}\right)}$$

(5.73)

To calculate settlement at a corner of the foundation, ·

$$\alpha' = 1$$

(5.74)

$$m' = \frac{L}{B}$$

(5.75)

and

$$n' = \frac{H}{B} \qquad (5.76)$$

The variations of F_1 and F_2 with m' and n' are given in Tables 5.12 and 5.13. Based on the work of Fox,[20] the variations of depth factor I_f for $v = 0.3, 0.4,$ and 0.5 and L/B are given in Figure 5.18. Note that I_f is not a function of H/B.

Due to the nonhomogeneous nature of a soil deposit, the magnitude of E_s may vary with depth. For that reason, Bowles[18] recommended

$$E_s = \frac{\sum E_{s(i)} \Delta z}{\bar{z}} \qquad (5.77)$$

where
$E_{s(I)}$ = soil modulus within the depth Δz
$\bar{z}$ = H or $5B$, whichever is smaller

Bowles[18] also recommended that

$$E_s = 500(N_{60} + 15) \text{ kN/m}^2 \qquad (5.78)$$

The elastic settlement of a rigid foundation can be estimated as

$$S_{e(\text{rigid})} \approx 0.93 S_{e(\text{flexible, center})} \qquad (5.79)$$

EXAMPLE 5.5

A rigid shallow foundation 1 m × 2 m is shown in Figure 5.19. Calculate the elastic settlement of the foundation.

Solution

We are given that $B = 1$ m and $L = 2$ m. Note that $\bar{z} = 5$ m $= 5B$. From equation (5.77),

$$E_s = \frac{\sum E_{s(i)} \Delta z}{\bar{z}} = \frac{(10,000)(2) + (8,000)(1) + (12,000)(2)}{5} = 10,400 \text{ kN/m}^2$$

For the center of the foundation,

$$\alpha' = 4$$

$$m' = \frac{L}{B} = \frac{2}{1} = 2$$

and

$$n' = \frac{H}{\left(\dfrac{B}{2}\right)} = \frac{5}{\left(\dfrac{1}{2}\right)} = 10$$

TABLE 5.12
Variation of F_1 with m' and n'

n'										m'										
	1.0	1.2	1.4	1.6	1.8	2.0	2.5	3.0	3.5	4.0	4.5	5.0	6.0	7.0	8.0	9.0	10.0	25.0	50.5	100.0
0.25	0.014	0.013	0.012	0.011	0.011	0.011	0.010	0.010	0.010	0.010	0.010	0.010	0.010	0.010	0.010	0.010	0.010	0.010	0.010	0.010
0.50	0.049	0.046	0.044	0.042	0.041	0.040	0.038	0.038	0.037	0.037	0.036	0.036	0.036	0.036	0.036	0.036	0.036	0.036	0.036	0.036
0.75	0.095	0.090	0.087	0.084	0.082	0.080	0.077	0.076	0.074	0.074	0.073	0.073	0.072	0.072	0.072	0.072	0.071	0.071	0.071	0.071
1.00	0.142	0.138	0.134	0.130	0.127	0.125	0.121	0.118	0.116	0.115	0.114	0.113	0.112	0.112	0.112	0.111	0.111	0.110	0.110	0.110
1.25	0.186	0.183	0.179	0.176	0.173	0.170	0.165	0.161	0.158	0.157	0.155	0.154	0.153	0.152	0.152	0.151	0.151	0.150	0.150	0.150
1.50	0.224	0.224	0.222	0.219	0.216	0.213	0.207	0.203	0.199	0.197	0.195	0.194	0.192	0.191	0.190	0.190	0.189	0.188	0.188	0.188
1.75	0.257	0.259	0.259	0.258	0.255	0.253	0.247	0.242	0.238	0.235	0.233	0.232	0.229	0.228	0.227	0.226	0.225	0.223	0.223	0.223
2.00	0.285	0.290	0.292	0.292	0.291	0.289	0.284	0.279	0.275	0.271	0.269	0.267	0.264	0.262	0.261	0.260	0.259	0.257	0.256	0.256
2.25	0.309	0.317	0.321	0.323	0.323	0.322	0.317	0.313	0.308	0.305	0.302	0.300	0.296	0.294	0.293	0.291	0.291	0.287	0.287	0.287
2.50	0.330	0.341	0.347	0.350	0.351	0.351	0.348	0.344	0.340	0.336	0.333	0.331	0.327	0.324	0.322	0.321	0.320	0.316	0.315	0.315
2.75	0.348	0.361	0.369	0.374	0.377	0.378	0.377	0.373	0.369	0.365	0.362	0.359	0.355	0.352	0.350	0.348	0.347	0.343	0.342	0.342
3.00	0.363	0.379	0.389	0.396	0.400	0.402	0.402	0.400	0.396	0.392	0.389	0.386	0.382	0.378	0.376	0.374	0.373	0.368	0.367	0.367
3.25	0.376	0.394	0.406	0.415	0.420	0.423	0.426	0.424	0.421	0.418	0.415	0.412	0.407	0.403	0.401	0.399	0.397	0.391	0.390	0.390
3.50	0.388	0.408	0.422	0.431	0.438	0.442	0.447	0.447	0.444	0.441	0.438	0.435	0.430	0.427	0.424	0.421	0.420	0.413	0.412	0.411
3.75	0.399	0.420	0.436	0.447	0.454	0.460	0.467	0.458	0.466	0.464	0.461	0.458	0.453	0.449	0.446	0.443	0.441	0.433	0.432	0.432
4.00	0.408	0.431	0.448	0.460	0.469	0.476	0.484	0.487	0.486	0.484	0.482	0.479	0.474	0.470	0.466	0.464	0.462	0.453	0.451	0.451
4.25	0.417	0.440	0.458	0.472	0.481	0.484	0.495	0.514	0.515	0.515	0.516	0.496	0.484	0.473	0.471	0.471	0.470	0.468	0.462	0.460
4.50	0.424	0.450	0.469	0.484	0.495	0.503	0.516	0.521	0.522	0.522	0.520	0.517	0.513	0.508	0.505	0.502	0.499	0.489	0.487	0.487
4.75	0.431	0.458	0.478	0.494	0.506	0.515	0.530	0.536	0.539	0.539	0.537	0.535	0.530	0.526	0.523	0.519	0.517	0.506	0.504	0.503
5.00	0.437	0.465	0.487	0.503	0.516	0.526	0.543	0.551	0.554	0.554	0.554	0.552	0.548	0.543	0.540	0.536	0.534	0.522	0.519	0.519
5.25	0.443	0.472	0.494	0.512	0.526	0.537	0.555	0.564	0.568	0.569	0.569	0.568	0.564	0.560	0.556	0.553	0.550	0.537	0.534	0.534
5.50	0.448	0.478	0.501	0.520	0.534	0.546	0.566	0.576	0.581	0.584	0.584	0.583	0.579	0.575	0.571	0.568	0.565	0.551	0.549	0.548
5.75	0.453	0.483	0.508	0.527	0.542	0.555	0.576	0.588	0.594	0.597	0.597	0.597	0.594	0.590	0.586	0.583	0.580	0.565	0.562	0.562
6.00	0.457	0.489	0.514	0.534	0.550	0.563	0.585	0.598	0.606	0.609	0.611	0.610	0.608	0.604	0.601	0.598	0.595	0.579	0.576	0.575
6.25	0.461	0.493	0.519	0.540	0.557	0.570	0.594	0.609	0.617	0.621	0.623	0.623	0.621	0.618	0.615	0.611	0.608	0.592	0.589	0.588

6.50	0.600	0.601	0.605	0.622	0.625	0.628	0.631	0.634	0.635	0.635	0.632	0.627	0.618	0.603	0.577	0.563	0.546	0.524	0.498	0.465
6.75	0.612	0.613	0.617	0.634	0.637	0.641	0.644	0.646	0.647	0.646	0.643	0.637	0.627	0.610	0.584	0.569	0.551	0.529	0.502	0.468
7.00	0.623	0.624	0.628	0.647	0.650	0.653	0.656	0.658	0.658	0.656	0.653	0.646	0.635	0.618	0.590	0.575	0.556	0.533	0.506	0.471
7.25	0.634	0.635	0.640	0.659	0.662	0.665	0.668	0.669	0.669	0.666	0.662	0.655	0.643	0.625	0.596	0.580	0.561	0.538	0.509	0.474
7.50	0.645	0.646	0.651	0.670	0.673	0.676	0.679	0.680	0.679	0.676	0.671	0.663	0.650	0.631	0.601	0.585	0.565	0.541	0.513	0.477
7.75	0.655	0.656	0.661	0.681	0.684	0.687	0.689	0.690	0.688	0.685	0.680	0.671	0.658	0.637	0.606	0.589	0.569	0.545	0.516	0.480
8.00	0.665	0.666	0.672	0.692	0.695	0.698	0.700	0.700	0.697	0.694	0.688	0.678	0.664	0.643	0.611	0.594	0.573	0.549	0.519	0.482
8.25	0.675	0.676	0.682	0.703	0.705	0.708	0.710	0.710	0.706	0.702	0.695	0.685	0.670	0.648	0.615	0.598	0.577	0.552	0.522	0.485
8.50	0.684	0.686	0.692	0.713	0.715	0.718	0.719	0.719	0.714	0.710	0.703	0.692	0.676	0.653	0.619	0.601	0.580	0.555	0.524	0.487
8.75	0.693	0.695	0.701	0.723	0.725	0.727	0.728	0.727	0.722	0.717	0.710	0.698	0.682	0.658	0.623	0.605	0.583	0.558	0.527	0.489
9.00	0.702	0.704	0.710	0.732	0.735	0.736	0.737	0.736	0.730	0.725	0.716	0.705	0.687	0.663	0.627	0.609	0.587	0.560	0.529	0.491
9.25	0.711	0.713	0.719	0.742	0.744	0.745	0.746	0.744	0.737	0.731	0.723	0.710	0.693	0.667	0.631	0.612	0.589	0.563	0.531	0.493
9.50	0.719	0.721	0.728	0.751	0.753	0.754	0.754	0.752	0.744	0.738	0.719	0.716	0.697	0.671	0.634	0.615	0.592	0.565	0.533	0.495
9.75	0.727	0.729	0.737	0.759	0.761	0.762	0.762	0.759	0.751	0.744	0.735	0.721	0.702	0.675	0.638	0.618	0.595	0.568	0.536	0.496
10.00	0.735	0.738	0.745	0.768	0.770	0.770	0.770	0.766	0.758	0.750	0.740	0.726	0.707	0.679	0.641	0.621	0.597	0.570	0.537	0.498
20.00	0.957	0.965	0.982	0.977	0.969	0.959	0.945	0.925	0.896	0.878	0.858	0.830	0.797	0.756	0.702	0.677	0.647	0.614	0.575	0.529
50.00	1.261	1.279	1.265	1.146	1.125	1.100	1.070	1.034	0.989	0.962	0.931	0.895	0.853	0.803	0.740	0.711	0.678	0.640	0.598	0.548
100	1.499	1.489	1.408	1.209	1.182	1.150	1.114	1.072	1.020	0.990	0.956	0.918	0.872	0.819	0.753	0.722	0.688	0.649	0.605	0.555

TABLE 5.13

Variation of F_2 with m' and n'

| n' | \multicolumn{20}{c}{m'} |
	1	1.2	1.4	1.6	1.8	2	2.5	3	3.5	4	4.5	5	6	7	8	9	10	25	50	100
0.25	0.049	0.050	0.051	0.051	0.051	0.052	0.052	0.052	0.052	0.052	0.053	0.053	0.053	0.053	0.053	0.053	0.053	0.053	0.053	0.053
0.50	0.074	0.077	0.080	0.081	0.083	0.084	0.086	0.086	0.087	0.087	0.087	0.087	0.088	0.088	0.088	0.088	0.088	0.088	0.088	0.088
0.75	0.083	0.089	0.093	0.097	0.099	0.101	0.104	0.106	0.107	0.108	0.109	0.109	0.109	0.110	0.110	0.110	0.110	0.111	0.111	0.111
1.00	0.083	0.091	0.098	0.102	0.106	0.109	0.114	0.117	0.119	0.120	0.121	0.122	0.123	0.123	0.124	0.124	0.124	0.125	0.125	0.125
1.25	0.080	0.089	0.096	0.102	0.107	0.111	0.118	0.122	0.125	0.127	0.128	0.130	0.131	0.132	0.132	0.133	0.133	0.134	0.134	0.134
1.50	0.075	0.084	0.093	0.099	0.105	0.110	0.118	0.124	0.128	0.130	0.132	0.134	0.136	0.137	0.138	0.138	0.139	0.140	0.140	0.140
1.75	0.069	0.079	0.088	0.095	0.101	0.107	0.117	0.123	0.128	0.131	0.134	0.136	0.138	0.140	0.141	0.142	0.142	0.144	0.144	0.145
2.00	0.064	0.074	0.083	0.090	0.097	0.102	0.114	0.121	0.127	0.131	0.134	0.136	0.139	0.141	0.143	0.144	0.145	0.147	0.147	0.148
2.25	0.059	0.069	0.077	0.085	0.092	0.098	0.110	0.119	0.125	0.130	0.133	0.136	0.140	0.142	0.144	0.145	0.146	0.149	0.150	0.150
2.50	0.055	0.064	0.073	0.080	0.087	0.093	0.106	0.115	0.122	0.127	0.132	0.135	0.139	0.142	0.144	0.146	0.147	0.151	0.151	0.151
2.75	0.051	0.060	0.068	0.076	0.082	0.089	0.102	0.111	0.119	0.125	0.130	0.133	0.138	0.142	0.144	0.146	0.147	0.152	0.152	0.153
3.00	0.048	0.056	0.064	0.071	0.078	0.084	0.097	0.108	0.116	0.122	0.127	0.131	0.137	0.141	0.144	0.145	0.147	0.152	0.153	0.154
3.25	0.045	0.053	0.060	0.067	0.074	0.080	0.093	0.104	0.112	0.119	0.125	0.129	0.135	0.140	0.143	0.145	0.147	0.153	0.154	0.154
3.50	0.042	0.050	0.057	0.063	0.070	0.076	0.089	0.100	0.109	0.116	0.122	0.126	0.133	0.138	0.142	0.144	0.146	0.153	0.155	0.155
3.75	0.040	0.047	0.054	0.060	0.067	0.073	0.086	0.096	0.105	0.113	0.119	0.124	0.131	0.137	0.141	0.143	0.145	0.154	0.155	0.155
4.00	0.037	0.044	0.051	0.057	0.063	0.069	0.082	0.093	0.102	0.110	0.116	0.121	0.129	0.135	0.139	0.142	0.145	0.154	0.155	0.156
4.25	0.036	0.042	0.049	0.055	0.061	0.066	0.079	0.090	0.099	0.107	0.113	0.119	0.127	0.133	0.138	0.141	0.144	0.154	0.156	0.156
4.50	0.034	0.040	0.046	0.052	0.058	0.063	0.076	0.086	0.096	0.104	0.110	0.116	0.125	0.131	0.136	0.140	0.143	0.154	0.156	0.156
4.75	0.032	0.038	0.044	0.050	0.055	0.061	0.073	0.083	0.093	0.101	0.107	0.113	0.123	0.130	0.135	0.139	0.142	0.154	0.156	0.157
5.00	0.031	0.036	0.042	0.048	0.053	0.058	0.070	0.080	0.090	0.098	0.105	0.111	0.120	0.128	0.133	0.137	0.140	0.154	0.156	0.157
5.25	0.029	0.035	0.040	0.046	0.051	0.056	0.067	0.078	0.087	0.095	0.102	0.108	0.118	0.126	0.131	0.136	0.139	0.154	0.156	0.157
5.50	0.028	0.033	0.039	0.044	0.049	0.054	0.065	0.075	0.084	0.092	0.099	0.106	0.116	0.124	0.130	0.134	0.138	0.154	0.156	0.157
5.75	0.027	0.032	0.037	0.042	0.047	0.052	0.063	0.073	0.082	0.090	0.097	0.103	0.113	0.122	0.128	0.133	0.136	0.154	0.157	0.157
6.00	0.026	0.031	0.036	0.040	0.045	0.05	0.060	0.070	0.079	0.087	0.094	0.101	0.111	0.120	0.126	0.131	0.135	0.153	0.157	0.157

6.25	0.025	0.030	0.034	0.039	0.044	0.048	0.058	0.068	0.077	0.085	0.092	0.098	0.109	0.118	0.124	0.129	0.134	0.153	0.157	0.158
6.50	0.024	0.029	0.033	0.038	0.042	0.046	0.056	0.066	0.075	0.083	0.090	0.096	0.107	0.116	0.122	0.128	0.132	0.153	0.157	0.158
6.75	0.023	0.028	0.032	0.036	0.041	0.045	0.055	0.064	0.073	0.080	0.087	0.094	0.105	0.114	0.121	0.126	0.131	0.153	0.157	0.158
7.00	0.022	0.027	0.031	0.035	0.039	0.043	0.053	0.062	0.071	0.078	0.085	0.092	0.103	0.112	0.119	0.125	0.129	0.152	0.157	0.158
7.25	0.022	0.026	0.030	0.034	0.038	0.042	0.051	0.060	0.069	0.076	0.083	0.090	0.101	0.110	0.117	0.123	0.128	0.152	0.157	0.158
7.50	0.021	0.025	0.029	0.033	0.037	0.041	0.050	0.059	0.067	0.074	0.081	0.088	0.099	0.108	0.115	0.121	0.126	0.152	0.156	0.158
7.75	0.020	0.024	0.028	0.032	0.036	0.039	0.048	0.057	0.065	0.072	0.079	0.086	0.097	0.106	0.114	0.120	0.125	0.151	0.156	0.158
8.00	0.020	0.023	0.027	0.031	0.035	0.038	0.047	0.055	0.063	0.071	0.077	0.084	0.095	0.104	0.112	0.118	0.124	0.151	0.156	0.158
8.25	0.019	0.023	0.026	0.030	0.034	0.037	0.046	0.054	0.062	0.069	0.076	0.082	0.093	0.102	0.110	0.117	0.122	0.150	0.156	0.158
8.50	0.018	0.022	0.026	0.029	0.033	0.036	0.045	0.053	0.060	0.067	0.074	0.080	0.091	0.101	0.108	0.115	0.121	0.150	0.156	0.158
8.75	0.018	0.021	0.025	0.028	0.032	0.035	0.043	0.051	0.059	0.066	0.072	0.078	0.089	0.099	0.107	0.114	0.119	0.150	0.156	0.158
9.00	0.017	0.021	0.024	0.028	0.031	0.034	0.042	0.050	0.057	0.064	0.071	0.077	0.880	0.097	0.105	0.112	0.118	0.149	0.156	0.158
9.25	0.017	0.020	0.024	0.027	0.030	0.033	0.041	0.049	0.056	0.063	0.069	0.075	0.086	0.096	0.104	0.110	0.116	0.149	0.156	0.158
9.50	0.017	0.020	0.023	0.026	0.029	0.033	0.040	0.048	0.055	0.061	0.068	0.074	0.085	0.094	0.102	0.109	0.115	0.148	0.156	0.158
9.75	0.016	0.019	0.023	0.026	0.029	0.032	0.039	0.047	0.054	0.060	0.066	0.072	0.083	0.092	0.100	0.107	0.113	0.148	0.156	0.158
10.00	0.016	0.019	0.022	0.025	0.028	0.031	0.038	0.046	0.052	0.059	0.065	0.071	0.082	0.091	0.099	0.106	0.112	0.147	0.156	0.158
20.00	0.008	0.010	0.011	0.013	0.014	0.016	0.020	0.024	0.027	0.031	0.035	0.039	0.046	0.053	0.059	0.065	0.071	0.124	0.148	0.156
50.00	0.003	0.004	0.004	0.005	0.006	0.006	0.008	0.010	0.011	0.013	0.014	0.016	0.019	0.022	0.025	0.028	0.031	0.071	0.113	0.142
100.00	0.002	0.002	0.002	0.003	0.003	0.003	0.004	0.005	0.006	0.006	0.007	0.008	0.010	0.011	0.013	0.014	0.016	0.039	0.071	0.113

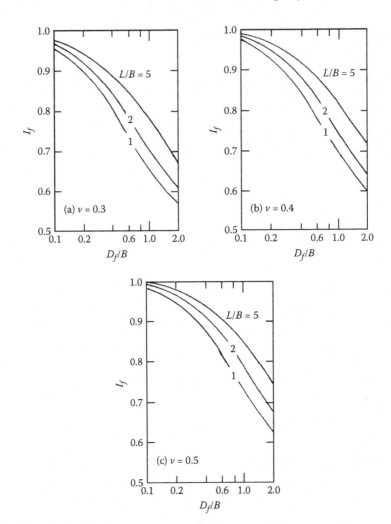

FIGURE 5.18 Variation of I_f with D_f/B. *Source:* Based on Fox, E. N. 1948. The mean elastic settlement of a uniformly loaded area at a depth below the ground surface, in *Proc., II Int. Conf. Soil Mech. Found. Eng.* 1:129; and Bowles, J. E. 1987. Elastic foundation settlement on sand deposits. *J. Geotech. Eng.*, ASCE, 113(8): 846.

From Tables 5.12 and 5.13, $F_1 = 0.641$ and $F_2 = 0.031$. From equation (5.64),

$$I_s = F_1 + \frac{2-v}{1-v}F_2 = 0.641 + \frac{2-0.3}{1-.03}(0.031) = 0.716$$

Again, $D_f/B = 1/1 = 1$; $L/B = 2$; and $v = 0.3$. From Figure 5.18, $I_f = 0.7$. Hence,

$$S_{e(\text{flexible})} = q(\alpha B')\frac{1-v^2}{E_s}I_s I_f = (200)\left(4\times\frac{1}{2}\right)\left(\frac{1-0.3^2}{10,400}\right)(0.716)(0.7) = 0.0175 \text{ m} = 17.5 \text{ mm}$$

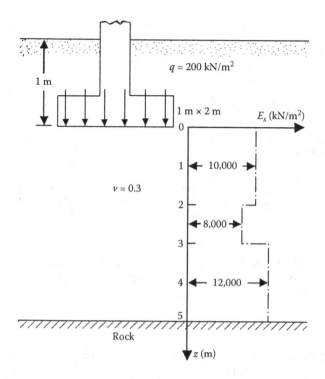

FIGURE 5.19 Elastic settlement below the center of a foundation.

Since the foundation is rigid, from equation (5.79) we obtain

$$S_{e(\text{rigid})} = (0.93)(17.5) = \textbf{16.3 mm}$$

5.4.7 ANALYSIS OF MAYNE AND POULOS BASED ON THE THEORY OF ELASTICITY—FOUNDATIONS ON GRANULAR SOIL

Mayne and Poulos[21] presented an improved formula for calculating the elastic settlement of foundations. The formula takes into account the rigidity of the foundation, the depth of embedment of the foundation, the increase in the modulus of elasticity of the soil with depth, and the location of rigid layers at a limited depth. To use Mayne and Poulos' equation, one needs to determine the equivalent diameter B_e of a rectangular foundation, or

$$B_e = \sqrt{\frac{4BL}{\pi}} \tag{5.80}$$

where
B = width of foundation
L = length of foundation

For circular foundations

$$B_e = B \tag{5.81}$$

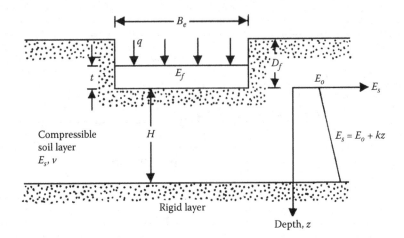

FIGURE 5.20 Mayne and Poulos' procedure for settlement calculation. *Source:* Mayne, P. W., and H. G. Poulos. 1999. Approximate displacement influence factors for elastic shallow foundations. *J. Geotech. Geoenviron. Eng.*, ASCE, 125(6): 453.

where
   B = diameter of foundation

Figure 5.20 shows a foundation with an equivalent diameter B_e located at a depth of D_f below the ground surface. Let the thickness of the foundation be t and the modulus of elasticity of the foundation material E_f. A rigid layer is located at a depth H below the bottom of the foundation. The modulus of elasticity of the compressible soil layer can be given as

$$E_s = E_o + kz \tag{5.82}$$

With the preceding parameters defined, the elastic settlement below the center of the foundation is

$$S_e = \frac{q B_e I_G I_R I_E}{E_o}(1-v^2) \tag{5.83}$$

where

I_G = influence factor for the variation of E_s with depth $= f\left(\beta = \dfrac{E_o}{kB_e}, \dfrac{H}{B_e}\right)$

I_R = foundation rigidity correction factor
I_E = foundation embedment correction factor

Figure 5.21 shows the variation of I_G with $\beta = E_o/kB_e$ and H/B_e. The foundation rigidity correction factor can be expressed as

$$I_R = \frac{\pi}{4} + \cfrac{1}{4.6 + 10\left(\cfrac{E_f}{E_o + \dfrac{B_e}{2}k}\right)\left(\dfrac{2t}{B_e}\right)^3} \tag{5.84}$$

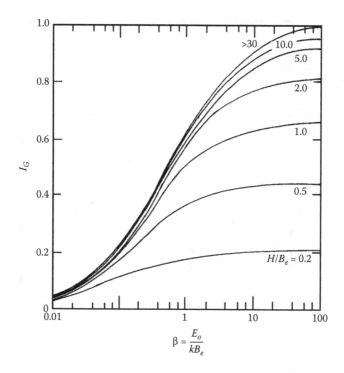

FIGURE 5.21 Variation of I_G with β.

Similarly, the embedment correction factor is

$$I_E = 1 - \frac{1}{3.5\exp(1.22v - 0.4)\left(\dfrac{B_e}{D_f} + 1.6\right)} \tag{5.85}$$

Figures 5.22 and 5.23 show the variation of I_R with I_E with the terms expressed in equations (5.84) and (5.85).

It is the opinion of the author that, if an average value of N_{60} within a zone of $3B$ to $4B$ below the foundation is determined, it can be used to estimate an average value of E_s and the magnitude of k can be assumed to be zero.

EXAMPLE 5.6

For a shallow foundation supported by silty clay, as shown in Figure 5.20, given:

Length $L = 1.5$ m
Width $B = 1$ m
Depth of foundation $D_f = 1$ m
Thickness of foundation $t = 0.23$ m
Net load per unit area $q = 190$ kN/m²
$E_f = 15 \times 10^6$ kN/m²

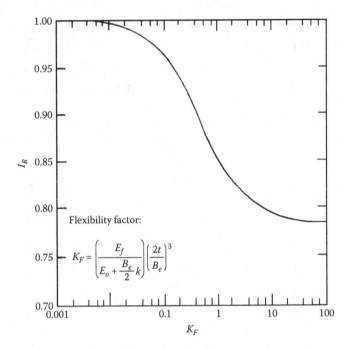

FIGURE 5.22 Variation of I_R.

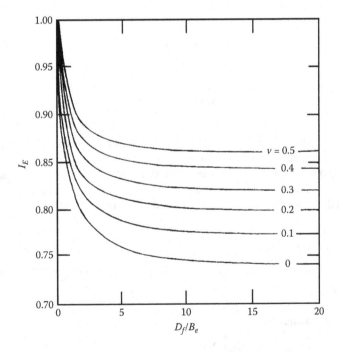

FIGURE 5.23 Variation of I_E.

The silty clay soil has the following properties:

$H = 2$ m

$v = 0.3$

$E_o = 9000$ kN/m²

$k = 500$ kN/m²

Estimate the elastic settlement of the foundation.

Solution

From equation (5.80), the equivalent diameter is

$$B_e = \sqrt{\frac{4BL}{\pi}} = \sqrt{\frac{(4)(1.5)(1)}{\pi}} = 1.38 \text{ m}$$

So,

$$\beta = \frac{E_o}{kB_e} = \frac{9000}{(500)(1.38)} = 13.04$$

and

$$\frac{H}{B_e} = \frac{2}{1.38} = 1.45$$

From Figure 5.21, for $\beta = 13.04$ and $H/B_e = 1.45$, the value of $I_G \approx 0.74$. From equation (5.84),

$$I_R = \frac{\pi}{4} + \frac{1}{4.6+10\left(\frac{E_f}{E_o+\frac{B_e}{2}k}\right)\left(\frac{2t}{B_e}\right)^3} = \frac{\pi}{4} + \frac{1}{4.6+10\left(\frac{15\times10^6}{9000+\left(\frac{1.38}{2}\right)(500)}\right)\left[\frac{(2)(0.23)}{1.38}\right]^3} = 0.787$$

From equation (5.85),

$$I_E = 1 - \frac{1}{3.5\exp(1.22v-0.4)\left(\frac{B_e}{D_f}+1.6\right)} = 1 - \frac{1}{3.5\exp[(1.22)(0.3)-0.4)]\left(\frac{1.38}{1}+1.6\right)} = 0.907$$

From equation (5.83),

$$S_e = \frac{qB_e I_G I_R I_E}{E_o}(1-v^2)$$

So, with $q = 190$ kN/m², it follows that

$$S_e = \frac{(190)(1.38)(0.74)(0.787)(0.907)}{9000}(1-0.3^2) = 0.014 \text{ m} \approx \textbf{14 mm}$$

5.4.8 ELASTIC SETTLEMENT OF FOUNDATIONS ON GRANULAR SOIL—ITERATION PROCEDURE

Berardi and Lancellotta[22] proposed a method to estimate the elastic settlement that takes into account the variation of the modulus of elasticity of soil with the strain

TABLE 5.14

Variation of I_F

L/B	Depth of Influence H_i/B			
	0.5	1.0	1.5	2.0
1	0.35	0.56	0.63	0.69
2	0.39	0.65	0.76	0.88
3	0.4	0.67	0.81	0.96
5	0.41	0.68	0.84	0.89
10	0.42	0.71	0.89	1.06

level. This method is also described by Berardi et al.[23] According to this procedure,

$$S_e = I_F \frac{qB}{E_s} \tag{5.86}$$

where

I_F = influence factor for a rigid foundation

This is based on the work of Tsytovich.[24] The variation of I_F for $v = 0.15$ is given in Table 5.14.

Analytical and numerical evaluations have shown that, for circular and square foundations, the depth H_{25} below the foundation beyond which the residual settlement is about 25% of the surface settlement can be taken as $0.8B$ to $1.3B$. For strip foundations ($L/B \geq 10$), H_{25} is about 50%–70% more compared to that for square foundations. Thus, the depth of influence H_i can be taken to be H_{25}. The modulus of elasticity E_s in equation (5.86) can be evaluated as

$$E_s = K_E p_a \left(\frac{\sigma_o' + 0.5\Delta\sigma'}{p_a} \right)^{0.5} \tag{5.87}$$

where

p_a = atmospheric pressure

σ_o' and $\Delta\sigma'$ = effective overburden stress and net effective stress increase due to the foundation loading, respectively, at a depth $B/2$ below the foundation

K_E = nondimensional modulus number

Berardi and Lancellota[22] reanalyzed the performance of 130 structures found on predominantly silica sand as reported by Burland and Burbidge,[16] and they obtained the variation of K_E with relative density D_r at $S_e/B = 0.1\%$ and K_E at varying strain levels. Figure 5.24a and b show the average variation of K_E with D_r at $S_e/B = 0.1\%$ and $[K_{E(S_e/B)}/K_{E(S_e/B=0.1\%)}]$ with S_e/B.

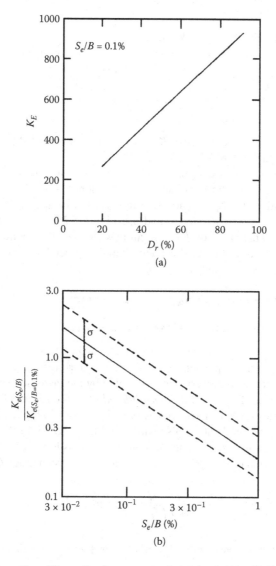

FIGURE 5.24 Berardi and Lancellota's recommended values: (a) variation of K_E with D_r; (b) variation of $\dfrac{K_e(S_e/B)}{K_e(S_e/B=0.1\%)}$ with S_e/B.

In order to estimate the elastic settlement of the foundation, an iterative procedure is suggested, which can be described as follows:

A. Determine the variation of the blow count from the standard penetration test N_{60} within the zone of influence, that is, H_{25}.

B. Determine the corrected blow count $(N_1)_{60}$ as

$$(N_1)_{60} = N_{60}\left(\frac{2}{1+0.01\sigma_o'}\right) \tag{5.88}$$

where

σ'_o = vertical effective stress (kN/m²)

C. Determine the average corrected blow count from standard penetration test $(\bar{N}_1)_{60}$ and, hence, the average relative density as

$$D_r = \left(\frac{\bar{N}_1}{60}\right)^{0.5} \tag{5.89}$$

D. With known D_r, determine $K_{E(S_e/B=0.1\%)}$ from Figure 5.24a and hence E_s from equation (5.87) for $S_e/B = 0.1\%$.

E. With the known value of E_s from step D, the magnitude of the elastic settlement S_e can be calculated from equation (5.86).

F. If the calculated S_e/B is not the same as the assumed S_e/B, then use the calculated S_e/B from step E and use Figure 5.24b to estimate a revised $K_{E(S_e/B)}$. This value of $K_{E(S_e/B)}$ can now be used in equations (5.87) and (5.86) to obtain a revised S_e. This iterative procedure can be continued until the assumed and calculated S_e are the same.

5.5 PRIMARY CONSOLIDATION SETTLEMENT

5.5.1 GENERAL PRINCIPLES OF CONSOLIDATION SETTLEMENT

As explained in section 5.1, consolidation settlement is a time-dependent process that occurs due to the expulsion of excess pore water pressure in saturated clayey soils below the groundwater table and is created by the increase in stress created by the foundation load. For *normally consolidated clay*, the nature of the variation of the void ratio e with vertical effective stress σ' is shown in Figure 5.25a. A similar plot for overconsolidated clay is also shown in Figure 5.25b. In this figure the *preconsolidation pressure* is σ'_c. The slope of the e versus log σ' plot for the normally consolidated portion of the soil is referred to as compression index C_c, or

$$C_c = \frac{e_1 - e_2}{\log\left(\frac{\sigma'_2}{\sigma'_1}\right)} \quad \text{(for } \sigma'_1 \leq \sigma'_c\text{)} \tag{5.90}$$

Similarly, the slope of the e versus log σ' plot for the overconsolidated portion of the clay is called the swell index C_s, or

$$C_s = \frac{e_3 - e_4}{\log\left(\frac{\sigma'_4}{\sigma'_3}\right)} \quad \text{(for } \sigma'_4 \leq \sigma'_c\text{)} \tag{5.91}$$

For normally consolidated clays, Terzaghi and Peck[25] gave a correlation for the compression index as

$$C_c = 0.009(LL - 10) \tag{5.92a}$$

where

LL = liquid limit

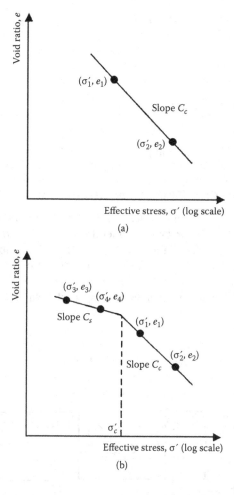

FIGURE 5.25 Nature of variation of void ratio with effective stress: (a) normally consolidated clay; (b) overconsolidated clay.

The preceding relation is reliable in the range of ±30% and should not be used for clays with sensitivity ratios greater than four.

Terzaghi and Peck[25] also gave a similar correlation for remolded clays:

$$C_c = 0.007(LL - 10) \tag{5.92b}$$

Several other correlations for the compression index with the basic index properties of soils have been made, and some of these are given below.[26]

$$C_c = 0.01 w_N \quad \text{(for Chicago clays)} \tag{5.93}$$

$$C_c = 0.0046(LL - 9) \quad \text{(for Brazilian clays)} \tag{5.94}$$

$$C_c = 1.21 + 1.055(e_o - 1.87) \quad \text{(for Motley clays, São Paulo city)} \tag{5.95}$$

$$C_c = 0.208e_o + 0.0083 \quad \text{(for Chicago clays)} \tag{5.96}$$

$$C_c = 0.0115w_N \tag{5.97}$$

where
w_N = natural moisture content in percent
e_o = in situ void ratio

The swell index C_s for a given soil is about 1/4 to 1/5 C_c.

5.5.2 Relationships for Primary Consolidation Settlement Calculation

Figure 5.26 shows a clay layer of thickness H_c. Let the initial void ratio before the construction of the foundation be e_o, and let the *average effective vertical stress* on the clay layer be σ'_o. The foundation located at a depth D_f is subjected to a net average pressure increase of q. This will result in an increase in the vertical stress in the soil. If the vertical stress increase at any point below the *center line* of the foundation is $\Delta\sigma$, the *average vertical stress increase* $\Delta\sigma_{av}$ in the clay layer can thus be given as

$$\Delta\sigma_{av} = \frac{1}{H_2 - H_1} \int_{z=H_1}^{z=H_2} (\Delta\sigma)dz \tag{5.98}$$

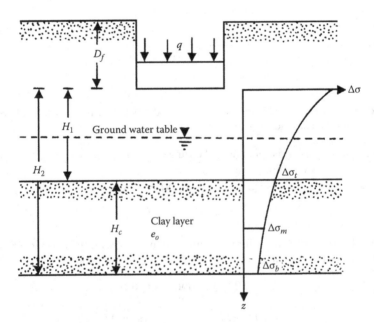

FIGURE 5.26 Primary consolidation settlement calculation.

The consolidation settlement S_c due to this average stress increase can be calculated as follows:

$$S_c = \frac{\Delta e}{1+e_o} = \frac{C_c H_c}{1+e_o} \log\left(\frac{\sigma'_o + \Delta\sigma_{av}}{\sigma'_o}\right)$$

(for normally consolidated clay, that is, $\sigma'_o = \sigma'_c$)   (5.99)

$$S_c = \frac{\Delta e}{1+e_o} = \frac{C_s H_c}{1+e_o} \log\left(\frac{\sigma'_o + \Delta\sigma_{av}}{\sigma'_o}\right)$$

(for overconsolidated clay, that is, $\sigma'_o + \Delta\sigma_{av} \geq \sigma'_c$)   (5.100)

$$S_c = \frac{\Delta e}{1+e_o} = \frac{C_s H_c}{1+e_o} \log\left(\frac{\sigma'_c}{\sigma'_o}\right) + \frac{C_c H_c}{1+e_o} \log\left(\frac{\sigma'_o + \Delta\sigma_{av}}{\sigma'_c}\right)$$

(for overconsolidated clay and $\sigma'_o < \sigma'_c < \sigma'_o + \Delta\sigma_{av}$)   (5.101)

where
Δe = change of void ratio due to primary consolidation

Equations (5.99), (5.100), and (5.101) can be used in two ways to calculate the primary consolidation settlement. They are:

Method A

According to this method, σ'_o is the in situ average of effective stress (that is, the effective stress at the middle of the clay layer). The magnitude of $\Delta\sigma_{av}$ can be calculated as (Figure 5.26)

$$\Delta\sigma_{av} = \tfrac{1}{6}(\Delta\sigma_t + 4\Delta\sigma_m + \Delta\sigma_b)$$

(5.102)

where
$\Delta\sigma_t$, $\Delta\sigma_m$, $\Delta\sigma_b$ = increase in stress at the top, middle, and bottom of the clay layer, respectively

The stress increase can be calculated by using the principles given previously in this chapter.

The average stress increase $\Delta\sigma_{av}$ from $z = 0$ to $z = H$ below the center of a uniformly loaded flexible rectangular area (Figure 5.27) was obtained by Griffiths[27] using the integration method, or

$$\Delta\sigma_{av} = q I_{av}$$

(5.103)

where

$$I_{av} = f\left(\frac{a}{H}, \frac{b}{H}\right)$$

(5.104)

a, b = half-length and half-width of the foundation

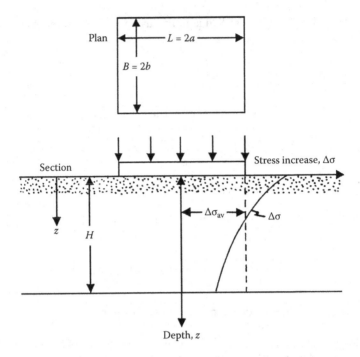

FIGURE 5.27 Average stress increase $\Delta\sigma_{av}$.

The variation of I_{av} is given in Figure 5.28 as a function of a/H and b/H. It is important to realize that I_{av} calculated by using this figure is for the case of average stress increase from $z = 0$ to $z = H$ (Figure 5.27). For calculating the average stress increase in a clay layer as shown in Figure 5.29,

$$I_{av(H_1/H_2)} = \frac{H_2 I_{av(H_2)} - H_1 I_{av(H_1)}}{H_2 - H_1}$$

where

$$I_{av(H_2)} = f\left(\frac{a}{H_2}, \frac{b}{H_2}\right)$$

$$I_{av(H_1)} = f\left(\frac{a}{H_1}, \frac{b}{H_1}\right)$$

$$H_2 - H_1 = H_c$$

So,

$$\Delta\sigma_{av} = q\left[\frac{H_2 I_{av(H_2)} - H_1 I_{av(H_1)}}{H_c}\right] \qquad (5.105)$$

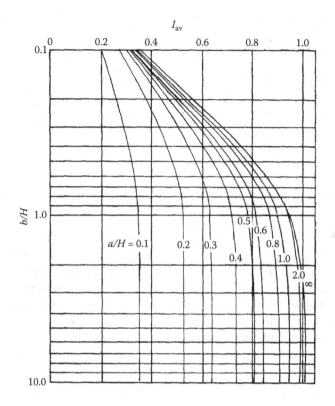

FIGURE 5.28 Variation of I_{av} with a/H and b/H.

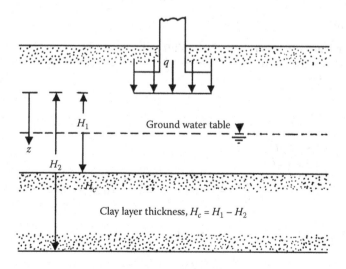

FIGURE 5.29 Average stress increase in a clay layer.

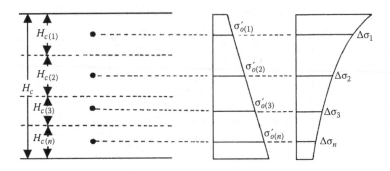

FIGURE 5.30 Consolidation settlement calculation using Method B.

Method B

In this method, a given clay layer can be divided into several thin layers having thicknesses of $H_{c(1)}, H_{c(2)}, \dots, H_{c(n)}$ (Figure 5.30). The in situ effective stresses at the middle of each layer are $\sigma'_{o(1)}, \sigma'_{o(2)}, \dots, \sigma'_{o(n)}$. The average stress increase for each layer can be approximated to be equal to the vertical stress increase at the middle of each soil layer [that is, $\Delta\sigma_{av(1)} \approx \Delta\sigma_1$, $\Delta\sigma_{av(2)} \approx \Delta\sigma_2$, $\dots$, $\Delta\sigma_{av(n)} \approx \Delta\sigma_n$]. Hence, the consolidation settlement of the entire layer can be calculated as

$$S_c = \sum_{i=1}^{i=n} \frac{\Delta e_i}{1 + e_{o(i)}} \tag{5.106}$$

EXAMPLE 5.7

Refer to Figure 5.31. Using Method A, determine the primary consolidation settlement of a foundation measuring 1.5 m × 3 m $(B \times L)$ in plan.

Solution

From equation (5.99) and given: $C_c = 0.27$; $H_c = 3$ m; $e_o = 0.92$,
$\sigma'_o = (1 + 1.5)(16.5) + (1.5)(17.8 - 9.81) + 3/2\,(18.2 - 9.81) = 65.82$ kN/m²

$$a = \frac{L}{2} = \frac{3}{2} = 1.5 \text{ m}$$

$$b = \frac{B}{2} = \frac{1.5}{2} = 0.75 \text{ m}$$

$H_1 = 1.5 + 1.5 = 3$ m
$H_2 = 1.5 + 1.5 + 3 = 6$ m

$$\frac{a}{H_1} = \frac{1.5}{3} = 0.5; \quad \frac{b}{H_1} = \frac{0.75}{3} = 0.25$$

From Figure 5.28, $I_{av(H_1)} = 0.54$. Similarly,

$$\frac{a}{H_2} = \frac{1.5}{6} = 0.25; \quad \frac{b}{H_2} = \frac{0.75}{6} = 0.125$$

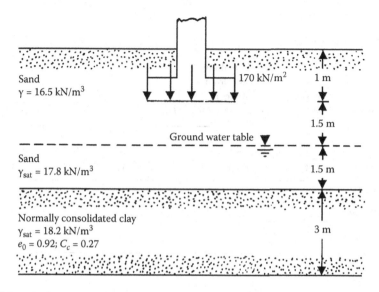

FIGURE 5.31 Consolidation settlement of a shallow foundation.

From Figure 5.28, $I_{av(H_2)} = 0.34$.
From equation (5.105),

$$\Delta\sigma_{av} = q\left[\frac{H_2 I_{av(H_2)} - H_1 I_{av(H_1)}}{H_c}\right] = 170\left[\frac{(6)(0.34) - (3)(0.54)}{3}\right] = 23.8 \text{ kN/m}^2$$

$$S_c = \frac{(0.27)(3)}{1+0.9}\log\left(\frac{65.82+23.8}{65.82}\right) = 0.057 \text{ m} = \textbf{57 mm}$$

EXAMPLE 5.8

Solve Example 5.7 by Method B. (Note: Divide the clay layer into three layers, each 1 m thick).

Solution

The following tables can now be prepared:

Calculation of σ'_o

Layer No.	Layer Thickness, H_i (m)	Depth to the Middle of Clay Layer (m)	σ'_o (kN/m²)
1	1	$1.0 + 1.5 + 1.5 + 0.5 = 4.5$	$(1+1.5)16.5 + (1.5)(17.8 - 9.81) + (0.5)$ $(18.2 - 9.81) = 57.43$
2	1	$4.5 + 1 = 5.5$	$57.43 + (1)(18.2 - 9.81) = 65.82$
3	1	$5.5 + 1 = 6.5$	$65.82 + (1)(18.2 - 9.81) = 74.21$

Calculation of $\Delta\sigma_{av}$

Layer No.	Layer Thickness H_i (m)	Depth to Middle of Layer from Bottom of Foundation, z (m)	L/B^a	z/B	$\dfrac{\Delta\sigma_{(av)_b}}{q}$	$\Delta\sigma_{av}{}^c$
1	1	3.5	2	2.33	0.16	27.2
2	1	4.5	2	3.0	0.095	16.15
3	1	5.5	2	3.67	0.07	11.9

$^a B = 1.5$ m; $L = 3$ m

^{b} Table 5.3

$^c q = 170$ kN/m²

$$S_c = \sum \frac{C_c H_i}{1+e_o} \log\left(\frac{\sigma'_{o(i)} + \Delta\sigma_{av(i)}}{\sigma'_{o(i)}}\right)$$

$$= \frac{(0.27)(1)}{1+0.9}\left[\log\left(\frac{57.43+27.2}{57.43}\right) + \log\left(\frac{65.82+16.15}{65.82}\right) + \log\left(\frac{74.21+11.9}{74.21}\right)\right]$$

$$= (0.142)(0.168+0.096+0.065) = 0.047 \text{ m} = \textbf{47 mm}$$

5.5.3 Three-Dimensional Effect on Primary Consolidation Settlement

The procedure described in the preceding section is for one-dimensional consolidation and will provide a good estimation for a field case where the width of the foundation is large relative to the thickness of the compressible stratum H_c, and also when the compressible material lies between two stiffer soil layers. This is because the magnitude of horizontal strains is relatively less in the above cases.

In order to account for the three-dimensional effect, Skempton and Bjerrum[28] proposed a correction to the one-dimensional consolidation settlement for normally consolidated clays. This can be explained by referring to Figure 5.32, which shows a circularly loaded area (diameter = B) on a layer of normally consolidated clay of thickness H_c. Let the stress increases at a depth z under the center line of the loaded area be $\Delta\sigma_1$ (vertical) and $\Delta\sigma_3$ (lateral). The increase in pore water pressure due to the increase in stress Δu can be given as

$$\Delta u = \Delta\sigma_3 + A(\Delta\sigma_1 - \Delta\sigma_3) \tag{5.107}$$

where
A = pore water pressure parameter

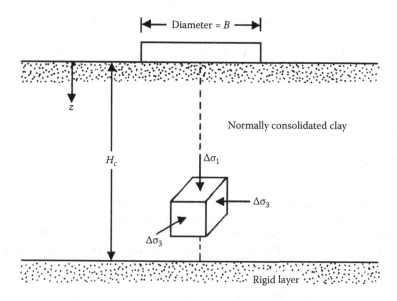

FIGURE 5.32 Three-dimensional effect on primary consolidation settlement (circular foundation of diameter B).

The consolidation settlement dS_c of an elemental soil layer of thickness dz is

$$dS_c = m_v \cdot \Delta u \cdot dz = \left[\frac{\Delta e}{(1+e_o)\Delta\sigma_1}\right](\Delta u)(dz) \qquad (5.108)$$

where

m_v = volume coefficient of compressibility
Δe = change in void ratio
e_o = initial void ratio

Hence,

$$S_c = \int dS_c = \int_0^{H_c} \left[\frac{\Delta e}{(1+e_o)\Delta\sigma_1}\right][\Delta\sigma_3 + A(\Delta\sigma_1 - \Delta\sigma_3)]dz$$

or

$$S_c = \int_0^{H_c} m_v \Delta\sigma_1 \left[A + \frac{\Delta\sigma_3}{\Delta\sigma_1}(1-A)\right] dz \qquad (5.109)$$

For conventional one-dimensional consolidation (section 5.5.1),

$$S_{c(oed)} = \int_0^{H_c} \frac{\Delta e}{1+e_o} dz = \int_0^{H_c} \frac{\Delta e}{\Delta\sigma_1(1+e_o)} \Delta\sigma_1 dz = \int_0^{H_c} m_v \Delta\sigma_1 dz \qquad (5.110)$$

From equations (5.109) and (5.110), the correction factor can be expressed as

$$\mu_{c(NC)} = \frac{S_c}{S_{c(oed)}} = \frac{\int_0^{H_c} m_v \Delta\sigma_1 [A + \frac{\Delta\sigma_3}{\Delta\sigma_1}(1-A)] \, dz}{\int_0^{H_c} m_v \Delta\sigma_1 \, dz} = A + (1-A) \frac{\int_0^{H_c} \Delta\sigma_3 \, dz}{\int_0^{H_c} \Delta\sigma_1 \, dz}$$

$$= A + (1-A)M_1 \tag{5.111}$$

where

$$M_1 = \frac{\int_0^{H_c} \Delta\sigma_3 \, dz}{\int_0^{H_c} \Delta\sigma_1 \, dz} \tag{5.112}$$

The variation of $\mu_{c(NC)}$ with A and H_c/B is shown in Figure 5.33.

In a similar manner, we can derive an expression for a uniformly loaded strip foundation of width B supported by a normally consolidated clay layer (Figure 5.34). Let $\Delta\sigma_1$, $\Delta\sigma_2$, and $\Delta\sigma_3$ be the increases in stress at a depth z below the center line of the foundation. For this condition, it can be shown that

$$\Delta u = \Delta\sigma_3 + \left[\frac{\sqrt{3}}{2}\left(A - \frac{1}{3}\right) + \frac{1}{2}\right](\Delta\sigma_1 - \Delta\sigma_3) \quad \text{(for } v = 0.5) \tag{5.113}$$

In a similar manner as equation (5.109),

$$S_c = \int_0^{H_c} m_v \Delta\sigma_1 \left[N + (1-N)\frac{\Delta\sigma_3}{\Delta\sigma_1}\right] dz \tag{5.114}$$

where

$$N = \frac{\sqrt{3}}{2}\left(A - \frac{1}{3}\right) + \frac{1}{2} \tag{5.115}$$

Thus,

$$\mu_{s(NC)} = \frac{S_c}{S_{c(oed)}} = \frac{\int_0^{H_c} m_v \Delta\sigma_1 \left[N + (1-N)\frac{\Delta\sigma_3}{\Delta\sigma_1}\right] dz}{\int_0^{H_c} m_v \Delta\sigma_1 dz} = N + (1-N)M_2 \tag{5.116}$$

where

$$M_2 = \frac{\int_0^{H_c} \Delta\sigma_3 dz}{\int_0^{H_c} \Delta\sigma_1 dz} \tag{5.117}$$

The plot of $\mu_{s(NC)}$ with A for varying values of H_c/B is shown in Figure 5.35.

Leonards[29] considered the correction factor $\mu_{c(OC)}$ for three-dimensional consolidation effect in the field for a circular foundation located over *overconsolidated clays*. Referring to Figure 5.36,

$$S_c = \mu_{c(OC)} S_{c(oed)} \tag{5.118}$$

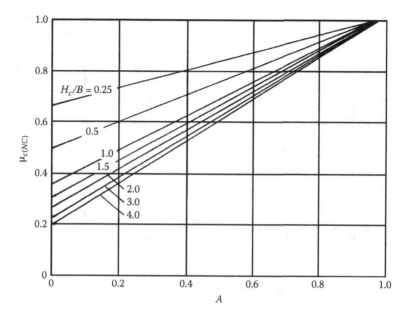

FIGURE 5.33 Variation of $\mu_{c(NC)}$ with A and H_c/B [equation (5.111)].

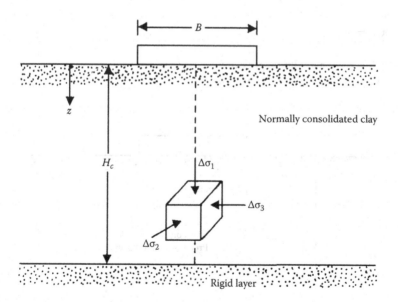

FIGURE 5.34 Three-dimensional effect on primary consolidation settlement (continuous foundation of width B).

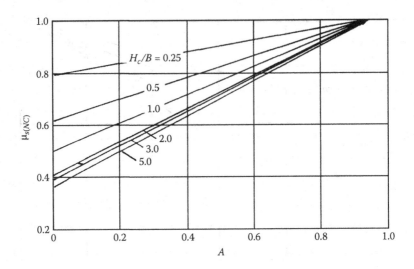

FIGURE 5.35 Variation of $\mu_{c(NC)}$ with A and H_c/B [equation (5.116)].

where

$$\mu_{c(OC)} = f\left(OCR, \frac{B}{H_c}\right) \tag{5.119}$$

$$OCR = \frac{\sigma'_c}{\sigma'_o} \tag{5.120}$$

σ'_c = preconsolidation pressure
σ'_o = present effective consolidation pressure

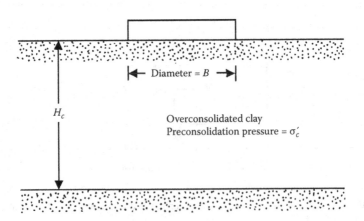

FIGURE 5.36 Three-dimensional effect on primary consolidation settlement of overconsolidated clays (circular foundation).

TABLE 5.15

Variation of $\mu_{c(OC)}$ with OCR and B/H_c

OCR	$\mu_{c(OC)}$		
	$B/H_c = 4.0$	$B/H_c = 1.0$	$B/H_c = 0.2$
1	1	1	1
2	0.986	0.957	0.929
3	0.972	0.914	0.842
4	0.964	0.871	0.771
5	0.950	0.829	0.707
6	0.943	0.800	0.643
7	0.929	0.757	0.586
8	0.914	0.729	0.529
9	0.900	0.700	0.493
10	0.886	0.671	0.457
11	0.871	0.643	0.429
12	0.864	0.629	0.414
13	0.857	0.614	0.400
14	0.850	0.607	0.386
15	0.843	0.600	0.371
16	0.843	0.600	0.357

The interpolated values of $\mu_{c(OC)}$ from the work of Leonards[29] are given in Table 5.15.

EXAMPLE 5.9

Refer to Example 5.7. Assume that the pore water pressure parameter A for the clay is 0.6. Considering the three-dimensional effect, estimate the consolidation settlement.

Solution

Note that equation (5.111) and Figure 5.33 are valid for only an axisymmetrical case; however, an approximate procedure can be adopted. Refer to Figure 5.37. If we assume that the load from the foundation spreads out along planes having slopes of 2V:1H, then the dimensions of the loaded area on the top of the clay layer are

$$B' = 1.5 + \tfrac{1}{2}(3) = 3 \text{ m}$$

$$L' = 3 + \tfrac{1}{2}(3) = 4.5 \text{ m}$$

The diameter of an equivalent circular area B_{eq} can be given as

$$\frac{\pi}{4} B_{eq}^2 = B'L'$$

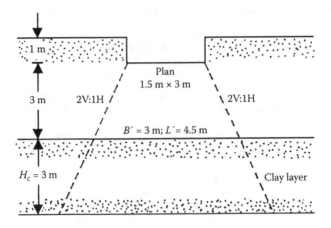

FIGURE 5.37 2 V: 1H load distribution under the foundation.

or

$$B_{eq}^2 = \sqrt{\frac{4}{\pi} B'L'} = \sqrt{\left(\frac{4}{\pi}\right)(3)(4.5)} \approx 4.15 \text{ m}$$

$$\frac{H_c}{B} = \frac{3}{4.15} = 0.723$$

From Figure 5.33, for $A = 0.6$ and $H_c/B = 0.723$, the magnitude of $\mu_{c(NC)} \approx 0.76$. So,

$$S_c = S_{c(oed)}\mu_{c(NC)} = (57)(0.76) = \textbf{43.3 mm}$$

5.6 SECONDARY CONSOLIDATION SETTLEMENT

5.6.1 SECONDARY COMPRESSION INDEX

Secondary consolidation follows the primary consolidation process and takes place under essentially constant effective stress as shown in Figure 5.38. The slope of the void ratio versus log-of-time plot is equal to C_α, or

$$C_\alpha = \text{secondary compression index} = \frac{\Delta e}{\log\left(\frac{t_2}{t_1}\right)} \tag{5.121}$$

The magnitude of the secondary compression index can vary widely, and some general ranges are as follows:

Overconsolidated clays (OCR >2 to 3)—>0.001
Organic soils—0.025 or more
Normally consolidated clays—0.004–0.025

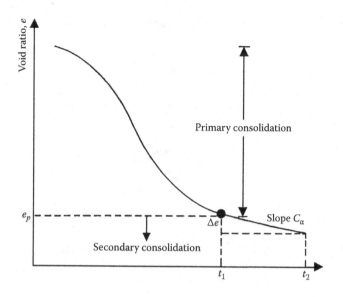

FIGURE 5.38 Secondary consolidation settlement.

5.6.2 SECONDARY CONSOLIDATION SETTLEMENT

The secondary consolidation settlement S_s can be calculated as

$$S_s = \frac{C_\alpha H_c}{1+e_p} \log\left(\frac{t_2}{t_1}\right) \qquad (5.122)$$

where

e_p = void ratio at the end of primary consolidation
t_2, t_1 = time

In a majority of cases, secondary consolidation is small compared to primary consolidation settlement. It can, however, be substantial for highly plastic clays and organic soils.

EXAMPLE 5.10

Refer to Example 5.7. Assume that the primary consolidation settlement is completed in 3 years. Also let $C_\alpha = 0.006$. Estimate the secondary consolidation settlement at the end of 10 years.

Solution

From equation (5.122),

$$S_s = \frac{C_\alpha H_c}{1+e_p} \log\left(\frac{t_2}{t_1}\right)$$

Given: $H_c = 3$ m, $C_\alpha = 0.006$, $t_2 = 10$ years, and $t_1 = 3$ years. From equation (5.90),

$$C_c = \frac{e_o - e_p}{\log\left(\dfrac{\sigma_2'}{\sigma_1'}\right)}$$

From Example 5.7, $\sigma_1' = 65.82$ kN/m², $\sigma_2' = 65.82 + 23.8 = 89.62$ kN/m², $C_c = 0.27$, $e_o = 0.92$. So,

$$0.27 = \frac{0.92 - e_p}{\log\left(\dfrac{89.62}{65.82}\right)}$$

$e_p = 0.884$,

$$S_s = \frac{(0.006)(3)}{1 + 0.884}\log\left(\frac{10}{3}\right) \approx 0.005 \text{ m} = \textbf{5 mm}$$

5.7 DIFFERENTIAL SETTLEMENT

5.7.1 GENERAL CONCEPT OF DIFFERENTIAL SETTLEMENT

In most instances, the subsoil is not homogeneous and the load carried by various shallow foundations of a given structure can vary widely. As a result, it is reasonable to expect varying degrees of settlement in different parts of a given building. The *differential settlement* of various parts of a building can lead to damage of the super-structure. Hence, it is important to define certain parameters to quantify differential settlement and develop limiting values for these parameters for desired safe performance of structures. Burland and Worth[30] summarized the important parameters relating to differential settlement. Figure 5.39 shows a structure in which various foundations at A, B, C, D, and E have gone through some settlement. The settlement at A is AA', and at B it is BB', ... Based on this figure the definitions of the various parameters follow:

S_T = total settlement of a given point
ΔS_T = difference between total settlement between any two parts
α = gradient between two successive points
β = angular distortion = $\frac{\Delta S_{T(ij)}}{l_{ij}}$ (Note: l_{ij} = distance between points i and j)
ω = tilt
Δ = relative deflection (that is, movement from a straight line joining two reference points)
$\frac{\Delta}{L}$ = deflection ratio

Since the 1950s, attempts have been made by various researchers and building codes to recommend allowable values for the above parameters. A summary of some of these recommendations is given in the following section.

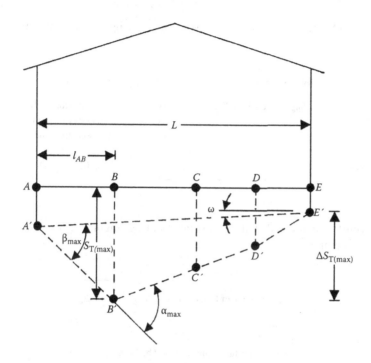

FIGURE 5.39 Definition of parameters for differential settlement.

5.7.2 LIMITING VALUE OF DIFFERENTIAL SETTLEMENT PARAMETERS

In 1956, Skempton and MacDonald[31] proposed the following limiting values for maximum settlement, maximum differential settlement, and maximum angular distortion to be used for building purposes:

Maximum settlement $S_{T(max)}$
   In sand—32 mm
   In clay—45 mm
Maximum differential settlement $\Delta S_{T(max)}$
   Isolated foundations in sand—51 mm
   Isolated foundations in clay—76 mm
   Raft in sand—51–76 mm
   Raft in clay—76–127 mm
Maximum angular distortion β_{max}—1/300

Based on experience, Polshin and Tokar[32] provided the allowable deflection ratios for buildings as a function of L/H (L = length; H = height of building), which are as follows:

$\Delta/L = 0.0003$ for $L/H \leq 2$
$\Delta/L = 0.001$ for $L/H = 8$

Shallow Foundations: Bearing Capacity and Settlement

The 1955 Soviet Code of Practice gives the following allowable values:

Building Type	L/H	Δ/L
Multistory buildings and civil dwellings	≤3	0.0003 (for sand)
		0.0004 (for clay)
	≥5	0.0005 (for sand)
		0.0007 (for clay)
One–story mills		0.001 (for sand and clay)

Bjerrum[33] recommended the following limiting angular distortions (β_{max}) for various structures:

Category of Potential Damage	β_{max}
Safe limit for flexible brick wall (L/H > 4)	1/150
Danger of structural damage to most buildings	1/150
Cracking of panel and brick walls	1/150
Visible tilting of high rigid buildings	1/250
First cracking of panel walls	1/300
Safe limit for no cracking of building	1/500
Danger to frames with diagonals	1/600

Grant et al.[34] correlated $S_{T(max)}$ and β_{max} for several buildings with the following results:

Soil Type	Foundation Type	Correlation
Clay	Isolated shallow foundation	$S_{T(max)}$ (mm) = 30,000 β_{max}
Clay	Raft	$S_{T(max)}$ (mm) = 35,000 β_{max}
Sand	Isolated shallow foundation	$S_{T(max)}$ (mm) = 15,000 β_{max}
Sand	Raft	$S_{T(max)}$ (mm) = 18,000 β_{max}

Using the above correlations, if the maximum allowable value of β_{max} is known, the magnitude of the allowable $S_{T(max)}$ can be calculated.

The European Committee for Standardization provided values for limiting values for serviceability limit states[35] and the maximum accepted foundation movements,[36] and these are given in Table 5.16.

TABLE 5.16

Recommendation of European Committee for Standardization on Differential Settlement Parameters

Item	Parameter	Magnitude	Comments
Limiting values for serviceability[35]	S_T	25 mm	Isolated shallow foundation
		50 mm	Raft foundation
	ΔS_T	5 mm	Frames with rigid cladding
		10 mm	Frames with flexible cladding
		20 mm	Open frames
	β	1/500	—
Maximum acceptable foundation movement[36]	S_T	50	Isolated shallow foundation
	ΔS_T	20	Isolated shallow foundation
	β	$\approx 1/500$	—

REFERENCES

1. Boussinesq, J. 1883. *Application des potentials a l'etude de l'equilibre et due mouvement des solides elastiques.* Paris: Gauthier-Villars.
2. Ahlvin, R. G., and H. H. Ulery. 1962. Tabulated values for determining the complete pattern of stresses, strains, and deflections beneath a uniform load on a homogeneous half space. *Highway Res. Rec., Bulletin* 342: 1.
3. Westergaard, H. M. 1938. A problem of elasticity suggested by a problem in soil mechanics: Soft material reinforced by numerous strong horizontal sheets, in *Contribution to the mechanics of solids, Stephen Timoshenko 60th anniversary vol.* New York: Macmillan.
4. Borowicka, H. 1936. Influence of rigidity of a circular foundation slab on the distribution of pressures over the contact surface, in *Proc., I Int. Conf. Soil Mech. Found. Eng.* 2: 144.
5. Trautmann, C. H., and F. H. Kulhawy. 1987. *CUFAD—A computer program for compression and uplift foundation analysis and design, report EL-4540-CCM,* 16. Palo Alto: Electric Power Research Institute.
6. Schmertmann, J. H. 1970. Static cone to compute settlement over sand. *J. Soil Mech. Found. Div.,* ASCE, 96(8): 1011.
7. Schmertmann, J. H., J. P. Hartman, and P. R. Brown. 1978. Improved strain influence factor diagrams. *J. Geotech. Eng. Div.,* ASCE, 104(8): 1131.
8. D'Appolonia, D. T., H. G. Poulos, and C. C. Ladd. 1971. Initial settlement of structures on clay. *J. Soil Mech. Found. Div.,* ASCE, 97(10): 1359.
9. Duncan, J. M., and A. L. Buchignani. 1976. *An engineering manual for settlement studies,* Department of Civil Engineering. Berkeley: University of California.
10. Janbu, N., L. Bjerrum, and B. Kjaernsli. 1956. *Veiledning ved losning av fundamenteringsoppgaver.* Oslo: Norwegian Geotechnical Institute Publication 16.
11. Christian, J. T., and W. D. Carrier III. 1978. Janbu, Bjerrum and Kjaernsli's chart reinterpreted. *Canadian Geotech. J.* 15(1): 124.
12. Terzaghi, K., and R. B. Peck. 1948. *Soil mechanics in engineering practice.* New York: Wiley.

13. Meyerhof, G. G. 1956. Penetration tests and bearing capacity of cohesionless soils. *J. Soil Mech. Found. Div.,* ASCE, 82(1): 1.

14. Meyerhof, G. G. 1965. Shallow foundations. *J. Soil Mech. Found. Div.,* ASCE, 91(2): 21.

15. Peck, R. B., and A R. S. S. Bazaraa. 1969. Discussion of paper by D'Appolonia et al. *J. Soil Mech. Found. Div.,* ASCE, 95(3): 305.

16. Burland, J. B., and M. C. Burbidge. 1985. Settlement of foundations on sand and gravel. *Proc., Institution of Civil Engineers* 78(1): 1325.

17. Eggestad, A. 1963. Deformation measurements below a model footing on the surface of dry sand, in *Proc. Eur Conf. Soil Mech. Found. Eng.* Weisbaden, W. Germany, 1: 223.

18. Bowles, J. E. 1987. Elastic foundation settlement on sand deposits. *J. Geotech. Eng.,* ASCE, 113(8): 846.

19. Steinbrenner, W. 1934. Tafeln zur setzungsberschnung. *Die Strasse* 1: 121.

20. Fox, E. N. 1948. The mean elastic settlement of a uniformly loaded area at a depth below the ground surface, in *Proc., II Int. Conf. Soil Mech. Found. Eng.* 1: 129.

21. Mayne, P. W., and H. G. Poulos. 1999. Approximate displacement influence factors for elastic shallow foundations. *J. Geotech. Geoenviron. Eng.,* ASCE 125(6): 453.

22. Berardi, R., and R. Lancellotta. 1991. Stiffness of granular soil from field performance. *Geotechnique* 41(1): 149.

23. Berardi, R., M. Jamiolkowski, and R. Lancellotta. 1991. Settlement of shallow foundations in sands: Selection of stiffness on the basis of penetration resistance. Geotechnical Engineering Congress 1991, *Geotech. Special Pub. 27,* ASCE, 185.

24. Tsytovich, N. A. 1951. Soil mechanics, ed. Stroitielstvo I. Archiketura, Moscow (in Russian).

25. Terzaghi, K., and R. B. Peck. 1967. *Soil mechanics in engineering practice,* 2nd Ed. New York: Wiley.

26. Azzouz, A. S., R. T. Krizek, and R. B. Corotis. 1976. Regression analysis of soil compressibility. *Soils and Found.* 16(2): 19.

27. Griffiths, D. V. 1984. A chart for estimating the average vertical stress increase in an elastic foundation below a uniformly loaded rectangular area. *Canadian Geotech. J.* 21(4): 710.

28. Skempton, A. W., and L. Bjerrum. 1957. A contribution to settlement analysis of foundations in clay. *Geotechnique* 7: 168.

29. Leonards, G. A. 1976. Estimating consolidation settlement of shallow foundations on overconsolidated clay. Transportation Research Board, Special Report 163, Washington, D.C.: 13.

30. Burland, J. B., and C. P. Worth. 1970. Allowable and differential settlement of structures, including damage and soil-structure interaction, in *Proc., Conf. on Settlement of Structures,* Cambridge University, U.K.: 11.

31. Skempton, A. W., and D. H. MacDonald, D. H. 1956. The allowable settlement of buildings, in *Proc., Institution of Civil Engineers,* 5, Part III: 727.

32. Polshin, D. E., and R. A. Tokar. 1957. Maximum allowable non-uniform settlement of structures, in *Proc., IV Int. Conf. Soil Mech. Found. Eng.,* London, 1: 402.

33. Bjerrum, L. 1963. Allowable settlement of structures, in *Proc., European Conf. Soil Mech. Found. Eng.,* Weisbaden Germany, 3: 135.

34. Grant, R., J. T. Christian, and E. H. Vanmarcke. 1974. Differential settlement of buildings. *J. Geotech. Eng. Div.,* ASCE, 100(9): 973.

35. European Committee for Standardization. 1994. Basis of design and actions on structures. Eurocode 1, Brussels, Belgium.

36. European Committee for Standardization. 1994. Geotechnical design, general rules— Part I. Eurocode 7, Brussels, Belgium.

6 Dynamic Bearing Capacity and Settlement

6.1 INTRODUCTION

Depending on the type of superstructure and the type of loading, a shallow foundation may be subjected to dynamic loading. The dynamic loading may be of various types, such as (a) monotonic loading with varying velocities, (b) earthquake loading, (c) cyclic loading, and (d) transient loading. The ultimate bearing capacity and settlement of shallow foundations subjected to dynamic loading are the topics of discussion of this chapter.

6.2 EFFECT OF LOAD VELOCITY ON ULTIMATE BEARING CAPACITY

The static ultimate bearing capacity of shallow foundations was discussed in Chapters 2, 3, and 4. Vesic et al.[1] conducted laboratory model tests to study the effect of the velocity of loading on the ultimate bearing capacity. These tests were conducted on a rigid rough circular model foundation having a diameter of 101.6 mm. The model foundation was placed on the *surface* of a dense sand layer. The velocity of loading to cause failure varied from about 25×10^{-5} mm/sec to 250 mm/sec. The tests were conducted in dry and submerged sand. From equation (2.82), for a surface foundation in sand subjected to vertical loading,

$$q_u = \tfrac{1}{2} \gamma B N_\gamma \lambda_{\gamma s}$$

or

$$N_\gamma \lambda_{\gamma s} = \frac{q_u}{\tfrac{1}{2} \gamma B} \tag{6.1}$$

where
  q_u = ultimate bearing capacity
  γ = effective unit weight of sand
  B = diameter of foundation
  N_γ = bearing capacity factor
  $\lambda_{\gamma s}$ = shape factor

The variation of $N_\gamma \lambda_{\gamma s}$ with the velocity of loading obtained in the study of Vesic et al.[1] is shown in Figure 6.1. It can be seen from this figure that, when the loading

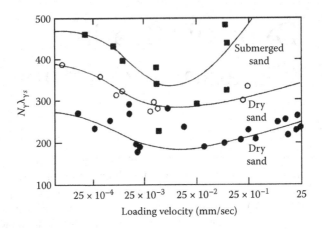

FIGURE 6.1 Variation of $N_\gamma \lambda_{\gamma s}$ with loading velocity. *Source:* After Vesic, A. S., D. C. Banks, and J. M. Woodward. 1965. An experimental study of dynamic bearing capacity of footings on sand, in *Proceedings, VI Int. Conf. Soil Mech. Found. Eng.*, Montreal, Canada, 2: 209.

velocity is between 25×10^{-3} mm/sec and 25×10^{-2} mm/sec, the ultimate bearing capacity reaches a minimum value. Vesic[2] suggested that the minimum value of q_u in *granular soil* can be obtained by using a soil friction angle of ϕ_{dy} instead of ϕ in the bearing capacity equation [equation (2.82)], which is conventionally obtained from laboratory tests, or

$$\phi_{dy} = \phi - 2° \qquad\qquad (6.2)$$

The above relationship is consistent with the findings of Whitman and Healy.[3] The increase in the ultimate bearing capacity when the loading velocity is very high is due to the fact that the soil particles in the failure zone do not always follow the path of least resistance, resulting in high shear strength of soil and thus ultimate bearing capacity.

Unlike in the case of sand, the undrained shear strength of saturated clay increases with the increase in the strain rate of loading. An excellent example can be obtained from the unconsolidated undrained triaxial tests conducted by Carroll[4] on buckshot clay. The tests were conducted with a chamber confining pressure ≈ 96 kN/m², and the moisture contents of the specimens were $33.5 \pm 0.2\%$. A summary of the test results follows:

Strain Rate (%/sec)	Undrained Cohesion c_u (kN/m²)
0.033	79.5
4.76	88.6
14.4	104
53.6	116.4
128	122.2
314 and 426	125.5

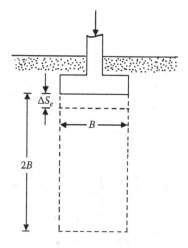

FIGURE 6.2 Strain rate definition under a foundation.

From the above data, it can be seen that $c_{u(\text{dynamic})}/c_{u(\text{static})}$ may be about 1.5. For a given foundation the strain rate $\dot{\varepsilon}$ can be approximated as (Figure 6.2)

$$\dot{\varepsilon} = \frac{1}{\Delta t}\left(\frac{\Delta S_e}{2B}\right) \tag{6.2}$$

where

    t = time

    S_e = settlement

So, if the undrained cohesion c_u ($\phi = 0$ condition) for a given soil at a given strain rate is known, this value can be used in equation (2.82) to calculate the ultimate bearing capacity.

6.3 ULTIMATE BEARING CAPACITY UNDER EARTHQUAKE LOADING

Richards et al.[5] proposed a bearing capacity theory for a *continuous* foundation supported by *granular soil* under earthquake loading. This theory assumes a simplified failure surface in soil at ultimate load. Figure 6.3a shows this failure surface under static conditions based on Coulomb's active and passive pressure wedges. Note that, in zone I, α_A is the angle that Coulomb's active wedge makes with the horizontal at failure:

$$\alpha_A = \phi + \tan^{-1}\left\{\frac{[\tan\phi(\tan\phi + \cot\phi)(1 + \tan\delta\cot\phi)]^{0.5} - \tan\phi}{1 + \tan\delta(\tan\phi + \cot\phi)}\right\} \tag{6.3}$$

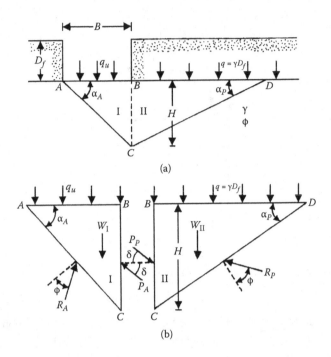

FIGURE 6.3 Bearing capacity of a continuous foundation on sand—static condition.

Similarly, in zone II, α_P is the angle that Coulomb's passive wedge makes with the horizontal at failure, or

$$\alpha_P = -\phi + \tan^{-1}\left\{\frac{[\tan\phi(\tan\phi+\cot\phi)(1+\tan\delta\cot\phi)]^{0.5}+\tan\phi}{1+\tan\delta(\tan\phi+\cot\phi)}\right\} \qquad (6.4)$$

where

    ϕ = soil friction angle
    δ = wall friction angle (*BC* in Figure 6.3a)

Considering a *unit length* of the foundation, Figure 6.3b shows the equilibrium analysis of wedges I and II. In this figure the following notations are used:

    P_A = Coulomb's active pressure
    P_P = Coulomb's passive pressure
    R_A = resultant of shear and normal forces along *AC*
    R_P = resultant of shear and normal forces along *CD*
  W_I, W_{II} = weight of wedges *ABC* and *BCD*, respectively

Now, if $\phi \neq 0$, $\gamma = 0$, and $q \neq 0$, then

$$q_u = q'_u$$

and

$$P_A \cos \delta = P_P \cos \delta \tag{6.5}$$

However,

$$P_A \cos \delta = q'_u K_A H \tag{6.6}$$

where

$H = \overline{BC}$

K_A = horizontal component of Coulomb's active earth pressure coefficient, or

$$K_A = \frac{\cos^2 \phi}{\cos \delta \left[1 + \sqrt{\frac{\sin(\phi + \delta) \sin \phi}{\cos \delta}} \right]^2} \tag{6.7}$$

Similarly,

$$P_P \cos \delta = q K_P H \tag{6.8}$$

where

K_P = horizontal component of Coulomb's passive earth pressure coefficient, or

$$K_P = \frac{\cos^2 \phi}{\cos \delta \left[1 - \sqrt{\frac{\sin(\phi - \delta) \sin \phi}{\cos \delta}} \right]^2} \tag{6.9}$$

Combining equations (6.5), (6.6), and (6.8),

$$q'_u = q \frac{K_P}{K_A} = q N_q \tag{6.10}$$

where

N_q = bearing capacity factor

Again, if $\phi \neq 0$, $\gamma \neq 0$, and $q = 0$, then $q_u = q''_u$:

$$P_A \cos \delta = q''_u H K_A + \tfrac{1}{2} \gamma H^2 K_A \tag{6.11}$$

Also,

$$P_P \cos \delta = \tfrac{1}{2} \gamma H^2 K_P \tag{6.12}$$

Equating the right-hand sides of equations (6.11) and (6.12),

$$q''_u H K_A + \tfrac{1}{2} \gamma H^2 K_A = \tfrac{1}{2} \gamma H^2 K_P$$

$$q''_u = \left[\tfrac{1}{2} \gamma H^2 (K_P - K_A) \right] \frac{1}{H K_A}$$

or

$$q_u'' = \frac{1}{2}\gamma H\left(\frac{K_P}{K_A} - 1\right)$$ (6.13)

However,

$$H = B\tan\alpha_A$$ (6.14)

Combining equations (6.13) and (6.14),

$$q_u'' = \frac{1}{2}\gamma B\tan\alpha_A\left(\frac{K_P}{K_A} - 1\right) = \frac{1}{2}\gamma BN_\gamma$$ (6.15)

where

$$N_\gamma = \text{bearing capacity factor} = \tan\alpha_A\left(\frac{K_P}{K_A} - 1\right)$$ (6.16)

If $\phi \neq 0$, $\gamma \neq 0$, and $q \neq 0$, using the superposition we can write

$$q_u = q_u' + q_u'' = qN_q + \frac{1}{2}\gamma BN_\gamma$$ (6.17)

Richards et al.[5] suggested that, in calculating the bearing capacity factors N_q and N_γ (which are functions of ϕ and δ), we may assume $\delta = \phi/2$. With this assumption, the variations of N_q and N_γ are given in Table 6.1.

It can also be shown that, for the $\phi = 0$ condition, if Coulomb's wedge analysis is performed, it will give a value of 6 for the bearing capacity factor N_c. For brevity we can assume

$$N_c = (N_q - 1)\cot\phi$$ (2.67)

Using equation (2.67) and the N_q values given in Table 6.1 the N_c values can be calculated, and these values are also shown in Table 6.1. Figure 6.4 shows the variations

TABLE 6.1
Variation of N_q, N_γ, and N_c (Assumption: $\delta = \phi/2$)

Soil Friction Angle ϕ (deg)	δ (deg)	N_q	N_γ	N_c
0	0	1	0	6
10	5	2.37	1.38	7.77
20	10	5.9	6.06	13.46
30	15	16.51	23.76	26.86
40	20	59.04	111.9	58.43

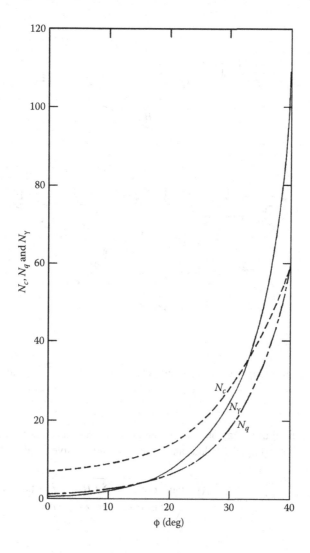

FIGURE 6.4 Variation of N_c, N_q, and N_γ with soil friction angle ϕ.

of the bearing capacity factors with soil friction angle ϕ. Thus, the ultimate bearing capacity q_u for a continuous foundation supported by a $c - \phi$ soil can be given as

$$q_u = cN_c + qN_q + \tfrac{1}{2}\gamma BN_\gamma \qquad (6.18)$$

The ultimate bearing capacity of a continuous foundation under earthquake loading can be evaluated in a manner similar to that for the static condition shown above. Figure 6.5 shows the wedge analysis for this condition for a foundation supported by

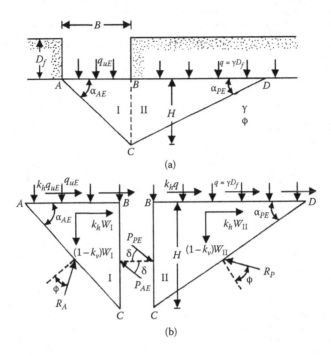

FIGURE 6.5 Bearing capacity of a continuous foundation on sand—earthquake condition.

granular soil. In Figure 6.5a note that α_{AE} and α_{PE} are, respectively, the angles that the Coulomb's failure wedges would make for active and passive conditions, or

$$\alpha_{AE} = \alpha + \tan^{-1}\left\{\frac{\sqrt{(1+\tan^2\alpha)[1+\tan(\delta+\theta)\cot\alpha]} - \tan\alpha}{1+\tan(\delta+\theta)(\tan\alpha+\cot\alpha)}\right\} \qquad (6.19)$$

and

$$\alpha_{PE} = -\alpha + \tan^{-1}\left\{\frac{\sqrt{(1+\tan^2\alpha)[1+\tan(\delta-\theta)\cot\alpha]} + \tan\alpha}{1+\tan(\delta+\theta)(\tan\alpha+\cot\alpha)}\right\} \qquad (6.20)$$

where
$$\alpha = \phi - \theta \qquad (6.21)$$
$$\theta = \tan^{-1}\frac{k_h}{1-k_v} \qquad (6.22)$$

k_h = horizontal coefficient of acceleration
k_v = vertical coefficient of acceleration

Figure 6.5b shows the equilibrium analysis of wedges I and II as shown in Figure 6.5a. As in the static analysis [similar to equation (6.17)],

$$q_{uE} = qN_{qE} + \tfrac{1}{2}\gamma BN_{\gamma E} \tag{6.23}$$

where

q_{uE} = ultimate bearing capacity
$N_{qE}, N_{\gamma E}$ = bearing capacity factors

Similar to equations (6.10) and (6.16),

$$N_{qE} = \frac{K_{PE}}{K_{AE}} \tag{6.24}$$

$$N_{\gamma E} = \tan \alpha_{AE} \left(\frac{K_{PE}}{K_{AE}} - 1 \right) \tag{6.25}$$

where

K_{AE}, K_{PE} = horizontal coefficients of active and passive earth pressure (under earthquake conditions), respectively, or

$$K_{AE} = \frac{\cos^2(\phi - \theta)}{\cos\theta \cos(\delta + \theta) \left[1 + \sqrt{\dfrac{\sin(\phi + \delta)\sin(\phi - \theta)}{\cos(\delta + \theta)}} \right]^2} \tag{6.26}$$

and

$$K_{PE} = \frac{\cos^2(\phi - \theta)}{\cos\theta \cos(\delta + \theta) \left[1 - \sqrt{\dfrac{\sin(\phi + \delta)\sin(\phi - \theta)}{\cos(\delta + \theta)}} \right]^2} \tag{6.27}$$

Using $\delta = \phi/2$ as before, the variations of K_{AE} and K_{PE} for various values of θ can be calculated. They can then be used to calculate the bearing capacity factors N_{qE} and $N_{\gamma E}$. Again, for a continuous foundation supported by a $c - \phi$ soil,

$$q_{uE} = cN_{cE} + qN_{qE} + \tfrac{1}{2}\gamma BN_{\gamma E} \tag{6.28}$$

where

N_{cE} = bearing capacity factor

The magnitude of N_{cE} can be approximated as

$$N_{cE} \approx (N_{qE} - 1)\cot\phi \tag{6.29}$$

Figures 6.6, 6.7, and 6.8 show the variations of $N_{\gamma E}/N_{\gamma}$, N_{qE}/N_q, and N_{cE}/N_c. These plots in combination with those given in Figure 6.4 can be used to estimate the ultimate bearing capacity of a continuous foundation q_{uE}.

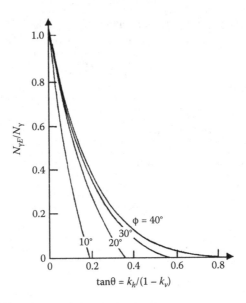

FIGURE 6.6 Variation of $N_{\gamma E}/N_\gamma$ with tan θ and ϕ. *Source:* After Richards, R., Jr., D. G. Elms, and M. Budhu. 1993. Seismic bearing capacity and settlement of foundations. *J. Geotech. Eng.*, ASCE, 119(4): 622.

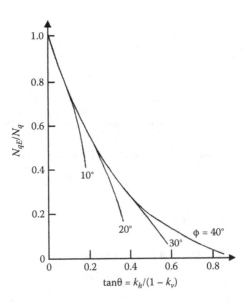

FIGURE 6.7 Variation of N_{qE}/N_q with tan θ and ϕ. *Source:* After Richards, R., Jr., D. G. Elms, and M. Budhu. 1993. Seismic bearing capacity and settlement of foundations. *J. Geotech. Eng.*, ASCE, 119(4): 622.

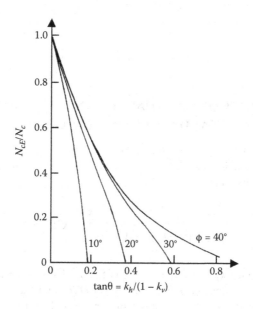

FIGURE 6.8 Variation of N_{cE}/N_c with tan θ and ϕ. *Source:* After Richards, R., Jr., D. G. Elms, and M. Budhu. 1993. Seismic bearing capacity and settlement of foundations. *J. Geotech. Eng.*, ASCE, 119(4): 622.

EXAMPLE 6.1

Consider a shallow continuous foundation. Given: $B = 1.5$ m; $D_f = 1$ m; $\gamma = 17$ kN/m³; $\phi = 25°$; $c = 30$ kN/m²; $k_h = 0.25$; $k_v = 0$. Estimate the ultimate bearing capacity q_{uE}.

Solution

From equation (6.28),

$$q_{uE} = cN_{cE} + qN_{qE} + \tfrac{1}{2}\gamma BN_{\gamma E}$$

For $\phi = 25°$, from Figure 6.4, $N_c \approx 20$, $N_q \approx 10$, and $N_\gamma \approx 14$. From Figures 6.6, 6.7, and 6.8, for tan $\theta = k_h/(1 - k_v) = 0.25/(1 - 0) = 0.25$,

$$\frac{N_{cE}}{N_c} = 0.44; \; N_{cE} = (0.44)(20) = 8.8$$

$$\frac{N_{qE}}{N_q} = 0.38; \; N_{qE} = (0.38)(10) = 3.8$$

$$\frac{N_{\gamma E}}{N_\gamma} = 0.13; \; N_{cE} = (0.13)(14) = 1.82$$

So,

$$q_{uE} = (30)(8.8) + (1 \times 17)(3.8) + \tfrac{1}{2}(17)(1.5)(1.82) = \textbf{351.8 kN/m}^2$$

6.4 SETTLEMENT OF FOUNDATION ON GRANULAR SOIL DUE TO EARTHQUAKE LOADING

Bearing capacity settlement of a foundation (supported by granular soil) during an earthquake takes place only when the critical acceleration ratio $k_h/(1 - k_v)$ reaches a certain critical value. Thus, if $k_v \approx 0$, then

$$\left(\frac{k_h}{1-k_v} \right)_{cr} \approx \left(\frac{k_h}{1-0} \right)_{cr} \approx k_h^*. \tag{6.30}$$

The critical value k_h^* is a function of the factor of safety FS taken over the ultimate static bearing capacity, embedment ratio D_f/B, and the soil friction angle ϕ. Richards et al.[5] developed this relationship, and it is shown in a graphical form in Figure 6.9. According to Richards et al.,[5] the settlement of a foundation during an earthquake can be given as

$$S_e = 0.174 \frac{V^2}{Ag} \left| \frac{k_h^*}{A} \right|^{-4} \tan \alpha_{AE} \tag{6.31}$$

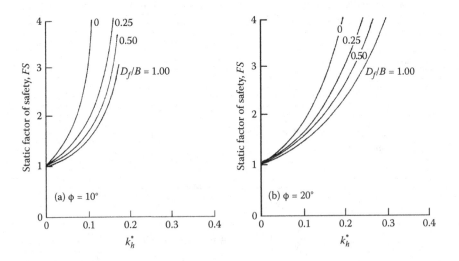

FIGURE 6.9 Critical acceleration k_h^* for incipient foundation settlement. *Source:* After Richards, R., Jr., D. G. Elms, and M. Budhu. 1993. Seismic bearing capacity and settlement of foundations. *J. Geotech. Eng.*, ASCE, 119(4): 622.

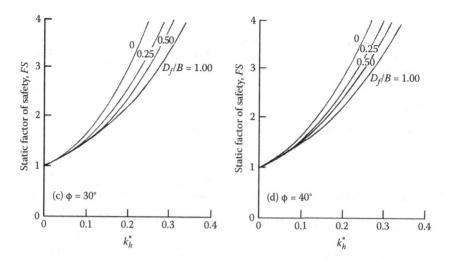

FIGURE 6.9 (Continued)

where

S_e = settlement
V = peak velocity of the design earthquake
A = peak acceleration coefficient of the design earthquake

The variations of $\tan \alpha_{AE}$ with k_h and ϕ are given in Figure 6.10.

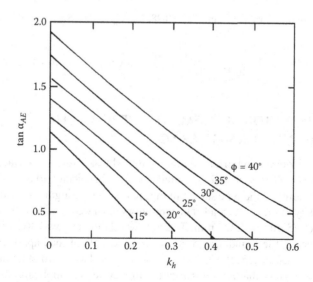

FIGURE 6.10 Variation of $\tan \alpha_{AE}$ with k_h and ϕ. *Source:* After Richards, R., Jr., D. G. Elms, and M. Budhu. 1993. Seismic bearing capacity and settlement of foundations. *J. Geotech. Eng.*, ASCE, 119(4): 622.

EXAMPLE 6.2

Consider a shallow foundation on granular soil with $B = 1.5$ m; $D_f = 1$ m; $\gamma = 16.5$ kN/m³; $\phi = 35°$. If the allowable bearing capacity is 304 kN/m², $A = 0.32$, and $V = 0.35$ m/s, determine the settlement the foundation may undergo.

Solution

From equation (6.17),

$$q_u = qN_q + \tfrac{1}{2}\gamma BN_\gamma$$

From Figure 6.4 for $\phi = 35°$, $N_q \approx 30$; $N_\gamma \approx 42$. So,

$$q_u = (1 \times 16.5)(30) + \tfrac{1}{2}(16.5)(1.5)(42) \approx 1015 \text{ kN/m}^2$$

Given $q_{all} = 340$ kN/m²,

$$FS = \frac{q_u}{q_{all}} = \frac{1015}{340} = 2.98$$

From Figure 6.9 for $FS = 2.98$ and $D_f/B = 1/1.5 = 0.67$, the magnitude of k_h^* is about 0.28. From equation (6.31),

$$S_e = 0.174 \frac{V^2}{Ag} \left| \frac{k_h^*}{A} \right|^{-4} \tan \alpha_{AE}$$

From Figure 6.10 for $\phi = 35°$ and $k_h^* = 0.28$, $\tan \alpha_{AE} \approx 0.95$. So,

$$S_e = (0.174) \frac{(0.35 \text{ m/s})^2}{(0.32)(9.81 \text{ m/s}^2)} \left| \frac{0.28}{0.32} \right|^{-4} (0.95) = 0.011 \text{ m} = \textbf{11 mm}$$

6.5 FOUNDATION SETTLEMENT DUE TO CYCLIC LOADING—GRANULAR SOIL

Raymond and Komos[6] reported laboratory model test results on surface continuous foundations ($D_f = 0$) supported by granular soil and subjected to a low-frequency (1 cps) cyclic loading of the type shown in Figure 6.11. In this figure, σ_d is the amplitude of the intensity of the cyclic load. The laboratory tests were conducted for foundation widths (B) of 75 mm and 228 mm. The unit weight of sand was 16.97 kN/m³. Since the settlement of the foundation S_e after the first cycle of load application was primarily due to the placement of the foundation rather than the foundation behavior, it was taken to be zero (that is, $S_e = 0$ after the first cycle load application). Figures 6.12 and 6.13 show the variation of S_e (after the first cycle) with the number of load cycles, N, and σ_d/q_u (q_u = ultimate static bearing capacity). Note that (a) for a given number of load cycles, the settlement increased with the increase in σ_d/q_u, and (b) for a given

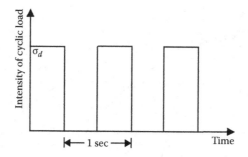

FIGURE 6.11 Cyclic load on a foundation.

σ_d/q_u, S_e increased with N. These load-settlement curves can be approximated by the relation (for $N = 2$ to 10^5)

$$S_e = \frac{a}{\dfrac{1}{\log N} - b}$$

(6.32)

where

$$a = -0.15125 + 0.0000693B^{1.18}\left(\frac{\sigma_d}{q_u} + 6.09\right)$$

(6.33)

$$b = -0.153579 + 0.0000363B^{0.821}\left(\frac{\sigma_d}{q_u} - 23.1\right)$$

(6.34)

In equations (6.33) and (6.34), B is in mm and σ_d/q_u is in percent.

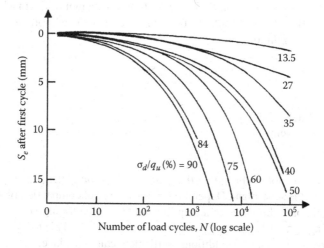

FIGURE 6.12 Variation of S_e (after first load cycle) with σ_d/q_u and N—$B = 75$ mm. *Source:* Raymond, G. P., and F. E. Komos. 1978. Repeated load testing of a model plane strain footing. *Canadian Geotech. J.* 15(2): 190.

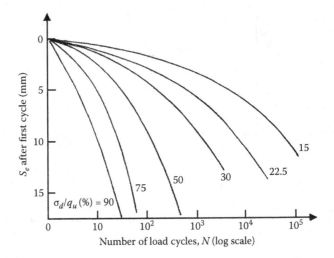

FIGURE 6.13 Variation of S_e (after first load cycle) with σ_d/q_u and N—$B = 228$ mm. *Source:* Raymond, G. P., and F. E. Komos. 1978. Repeated load testing of a model plane strain footing. *Canadian Geotech. J.* 15(2): 190.

Figures 6.14 and 6.15 show the contours of the variation of S_e with σ_d and N for $B = 75$ mm and 228 mm. Studies of this type are useful in designing railroad ties.

6.5.1 Settlement of Machine Foundations

Machine foundations subjected to *sinusoidal vertical vibration* (Figure 6.16) may undergo permanent settlement S_e. In Figure 6.16, the weight of the machine and the foundation is W and the diameter of the foundation is B. The impressed cyclic force Q is given by the relationship

$$Q = Q_o \sin \omega t \tag{6.35}$$

where
    Q_o = amplitude of the force
    ω = angular velocity
    t = time

Many investigators believe that the *peak acceleration* is the primary controlling parameter for the settlement. Depending on the degree of compaction of the granular soil, the solid particles come to an equilibrium condition for a given peak acceleration resulting in a settlement $S_{e(max)}$ as shown in Figure 6.17. This *threshold acceleration* must be attained before additional settlement can take place.

Brumund and Leonards[7] evaluated the settlement of *circular foundations* subjected to vertical sinusoidal loading by laboratory model tests. For this study the model foundation had a diameter of 101.6 mm, and 20–30 Ottawa sand compacted

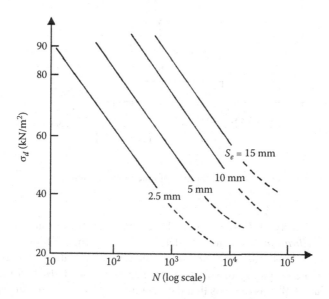

FIGURE 6.14 Contours of variation of S_e with σ_d and N—$B = 75$ mm. *Source:* Raymond, G. P., and F. E. Komos. 1978. Repeated load testing of a model plane strain footing. *Canadian Geotech. J.* 15(2): 190.

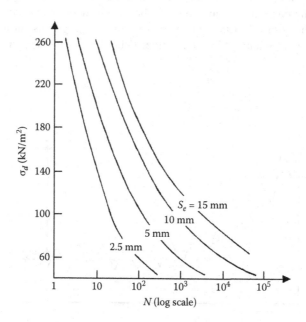

FIGURE 6.15 Contours of variation of S_e with σ_d and N—$B = 228$ mm. *Source:* Raymond, G. P., and F. E. Komos. 1978. Repeated load testing of a model plane strain footing. *Canadian Geotech. J.* 15(2): 190.

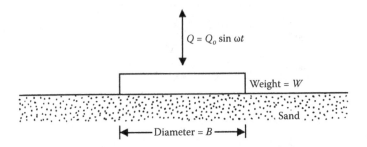

FIGURE 6.16 Sinusoidal vertical vibration of machine foundation.

at a relative density of 70% was used. Based on their study, it appears that *energy per cycle of vibration* can be used to determine $S_{e(max)}$. Figure 6.18 shows the variation of $S_{e(max)}$ versus peak acceleration for weights of foundation, $W = 217$ N, 327 N, and 436 N. The frequency of vibration was kept constant at 20 Hz for all tests. For a given value of W, it is obvious that the magnitude of S_e increases linearly with the peak acceleration level.

The maximum energy transmitted to the foundation per cycle of vibration can be theorized as follows. Figure 6.19 shows the schematic diagram of a lumped-parameter one-degree-of-freedom vibrating system for the machine foundation. The soil supporting the foundation has been taken to be equivalent to a *spring* and a *dashpot*. Let the spring constant be equal to k and the viscous damping constant of the dashpot be c. The spring constant k and the viscous damping constant c can be given by the following relationships (for further details see any soil dynamics text, for

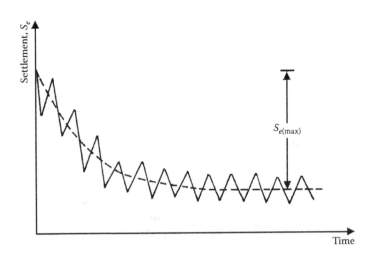

FIGURE 6.17 Settlement S_e with time due to cyclic load application.

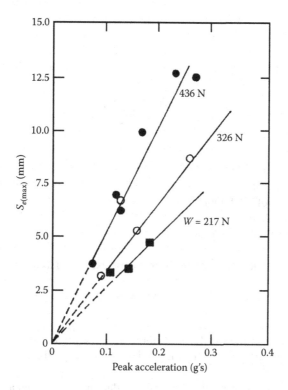

FIGURE 6.18 Variation of $S_{e(max)}$ with peak acceleration and weight of foundation. *Source:* Brumund, W. F., and G. A. Leonards. 1972. Subsidence of sand due to surface vibration. *J. Soil Mech. Found. Eng. Div.*, ASCE, 98(1): 27.

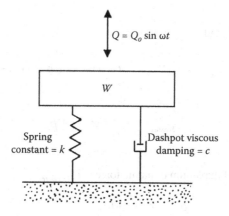

FIGURE 6.19 Lumped-parameter one-degree-of-freedom vibrating system.

example, Das[8]):

$$k = \frac{2GB}{1-v_s} \tag{6.36}$$

$$c = \frac{0.85}{1-v_s} B^2 \sqrt{\frac{G\gamma}{g}} \tag{6.37}$$

where
    G = shear modulus of the soil
    v_s = Poisson's ratio of the soil
    B = diameter of the foundation
    γ = unit weight of soil
    g = acceleration due to gravity

The vertical motion of the foundation can be expressed as

$$z = Z\cos(\omega t + \alpha) \tag{6.38}$$

where
    Z = amplitude of the steady-state vibration of the foundation
    α = phase angle by which the motion lags the impressed force

The *dynamic force* transmitted by the foundation can be given as

$$F_{dynamic} = kz + c\frac{dz}{dt} \tag{6.39}$$

Substituting equation (6.38) into equation (6.39) we obtain

$$F_{dynamic} = kZ\cos(\omega t + \alpha) - c\omega Z\sin(\omega t + \alpha)$$

Let $kZ = A\cos\beta$ and $c\omega Z = A\sin\beta$. So,

$$F_{dynamic} = A\cos\beta\,\cos(\omega t + \alpha) - A\sin\beta\,\sin(\omega t + \alpha)$$

or

$$F_{dynamic} = A\cos(\omega t + \alpha + \beta) \tag{6.40}$$

where
    A = magnitude of maximum dynamic force = $F_{dynamic(max)}$

$$= \sqrt{(A\cos\beta)^2 + (A\sin\beta)^2} = Z\sqrt{k^2 + (c\omega)^2} \tag{6.41}$$

The energy transmitted to the soil per cycle of vibration E_{tr} is

$$E_{tr} = \int F dz = F_{av} Z \tag{6.42}$$

where
  F = total contact force on soil
  F_{av} = average contact force on the soil

However,

$$F_{av} = \tfrac{1}{2}(F_{max} + F_{min}) \tag{6.43}$$

$$F_{max} = W + F_{\text{dynamics(max)}} \tag{6.44}$$

$$F_{max} = W - F_{\text{dynamics(max)}} \tag{6.45}$$

Combining equations (6.43), (6.44), and (6.45),

$$F_{av} = W \tag{6.46}$$

Hence, from equations (6.42) and (6.46),

$$E_{tr} = WZ \tag{6.47}$$

Figure 6.20 shows the experimental results of Brumund and Leonards,[7] which is a plot of $S_{e(max)}$ versus E_{tr}. The data include (a) a frequency range of 14–59.3 Hz,

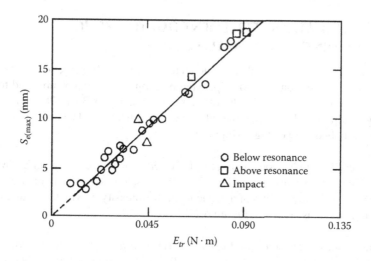

FIGURE 6.20 Plot of $S_{e(max)}$ versus E_{tr}. *Source:* Brumund, W. F., and G. A. Leonards. 1972. Subsidence of sand due to surface vibration. *J. Soil Mech. Found. Eng. Div.*, ASCE, 98(1): 27.

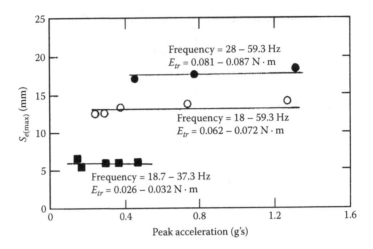

FIGURE 6.21 $S_{e(max)}$ versus peak acceleration for three levels of transmitted energy. *Source:* Brumund, W. F., and G. A. Leonards. 1972. Subsidence of sand due to surface vibration. *J. Soil Mech. Found. Eng. Div.*, ASCE, 98(1): 27.

(b) a range of W varying from $0.27q_u$ to $0.55q_u$, (q_u = static beaning capacity) and (c) the maximum downward dynamic force of $0.3W$ to $1.0W$. The results show that $S_{e(max)}$ increases linearly with E_{tr}. Figure 6.21 shows a plot of the experimental results of $S_{e(max)}$ against peak acceleration for different ranges of E_{tr}. This clearly demonstrates that, if the value of the transmitted energy is constant, the magnitude of $S_{e(max)}$ remains constant irrespective of the level of peak acceleration.

6.6 FOUNDATION SETTLEMENT DUE TO CYCLIC LOADING IN SATURATED CLAY

Das and Shin[9] provided small-scale model test results for the settlement of a continuous surface foundation ($D_f = 0$) supported by saturated clay and subjected to cyclic loading. For these tests, the width of the model foundation B was 76.2 mm, and the average undrained shear strength of the clay was 11.9 kN/m². The load to the foundation was applied in two stages (Figure 6.22):

Stage I—Application of a static load per unit area of $q_s = q_u/FS$ (where q_u = ultimate bearing capacity; FS = factor of safety) as shown in Figure 6.22a
Stage II—Application of a cyclic load, the intensity of which has an amplitude of σ_d as shown in Figure 6.22b

The frequency of the cyclic load was 1 Hz. Figure 6.22c shows the variation of the total load intensity on the foundation. Typical experimental plots obtained from these laboratory tests are shown by the dashed lines in Figure 6.23 ($FS = 3.33$; $\sigma_d/q_u = 4.38\%$, 9.38%, and 18.75%). It is important to note that S_e in this figure refers to the

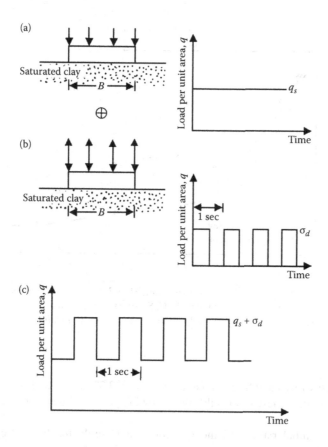

FIGURE 6.22 Load application sequence to observe foundation settlement in saturated clay due to cyclic loading based on laboratory model tests of Das and Shin. *Source:* Das, B. M., and E. C. Shin. 1996. Laboratory model tests for cyclic load-induced settlement of a strip foundation on a clayey soil. *Geotech. Geol. Eng.*, London, 14: 213.

settlement obtained due to cyclic load only (that is, after application of stage II load; Figure 6.22b). The general nature of these plots is shown in Figure 6.24. They consist of approximately three linear segments, and they are

1. An initial rapid settlement $S_{e(r)}$ (branch Oa).
2. A secondary settlement at a slower rate $S_{e(s)}$ (branch ab). The settlement practically ceases after application of $N = N_{cr}$ cycles of load.
3. For $N > N_{cr}$ cycles of loading, the settlement of the foundation due to cyclic load practically ceases (branch bc).

The linear approximations of S_e with number of load cycles N are shown in Figure 6.23 (solid lines). Hence the total settlement of the foundation is

$$S_{e(\text{max})} = S_{e(r)} + S_{e(s)} \qquad (6.48)$$

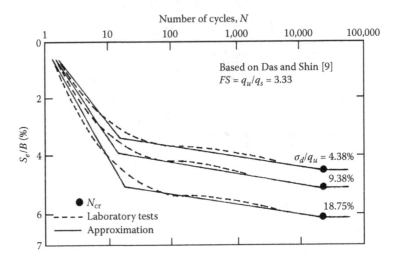

FIGURE 6.23 Typical plots of S_e/B versus N for $FS = 3.33$ and $\sigma_d/q_u = 4.38\%$, 9.38%, and 18.75%. *Source:* Based on Das, B. M., and E. C. Shin. 1996. Laboratory model tests for cyclic load-induced settlement of a strip foundation on a clayey soil. *Geotech. Geol. Eng.*, London, 14: 213.

The tests of Das and Shin[9] had a range of $FS = 3.33$ to 6.67 and $\sigma_d/q_u = 4.38\%$ to 18.75%. Based on these test results, the following general conclusions were drawn:

1. The initial rapid settlement is completed within the first 10 cycles of loading.
2. The magnitude of N_{cr} varied between 15,000 and 20,000 cycles. This is independent of FS and σ_d/q_u.
3. For a given FS, the magnitude of S_e increased with an increase of σ_d/q_u.
4. For a given σ_d/q_u, the magnitude of S_e increased with a decrease in FS.

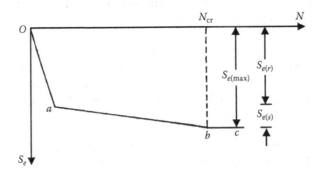

FIGURE 6.24 General nature of plot of S_e versus N for given values of FS and σ_d/q_u.

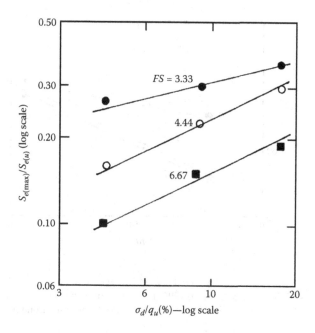

FIGURE 6.25 Results of laboratory model tests of Das and Shin—plot of $S_{e(max)}/S_{e(u)}$ versus σ_d/q_u. *Source:* Das, B. M., and E. C. Shin. 1996. Laboratory model tests for cyclic load-induced settlement of a strip foundation on a clayey soil. *Geotech. Geol. Eng.*, London, 14: 213.

Figure 6.25 shows a plot of $S_{e(max)}/S_{e(u)}$ versus σ_d/q_u for various values of *FS*. Note that $S_{e(u)}$ is the settlement of the foundation corresponding to the static ultimate bearing capacity. Similarly, Figure 6.26 is the plot of $S_{e(r)}/S_{e(max)}$ versus σ_d/q_u for various values of *FS*. From these plots it can be seen that

$$\frac{S_{e(max)}}{S_{e(u)}} = m_1 \underbrace{\left(\frac{\sigma_d}{q_u}\right)^{n_1}}_{\text{Fig. 6.25}}$$

and for any *FS* and σ_d/q_u (Figure 6.26), the limiting value of $S_{e(r)}$ may be about $0.8\, S_{e(max)}$.

6.7 SETTLEMENT DUE TO TRANSIENT LOAD ON FOUNDATION

A limited number of test results are available in the literature that relate to the evaluation of settlement of shallow foundations (supported by sand and clay) subjected to transient loading. The findings of these tests are discussed in this section.

Cunny and Sloan[10] conducted several model tests on square surface foundations ($D_f = 0$) to observe the settlement when the foundations were subjected to transient

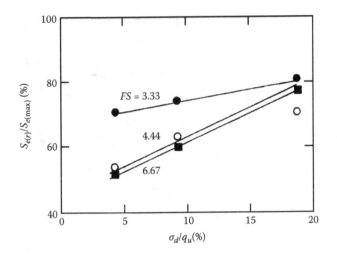

FIGURE 6.26 Results of laboratory model tests of Das and Shin—plot of $S_{e(r)}/S_{e(max)}$ versus σ_d/q_u. *Source:* Das, B. M., and E. C. Shin. 1996. Laboratory model tests for cyclic load-induced settlement of a strip foundation on a clayey soil. *Geotech. Geol. Eng.*, London, 14: 213.

loading. The nature of variation of the transient load with time used for this study is shown in Figure 6.27. Tables 6.2 and 6.3 show the results of these tests conducted in sand and clay, respectively. Other details of the tests are as follows:

Tests in Sand (Table 6.2)

Dry unit weight $\gamma = 16.26$ kN/m³
Relative density of compaction = 96%
Triaxial angle of friction = 32°

Tests in Clay (Table 6.3)
Compacted moist unit weight = 14.79 – 15.47 kN/m³
Moisture content = 22.5 ± 1.7%
Angle of friction (undrained triaxial test) = 4°
Cohesion (undrained triaxial test) = 115 kN/m²

For all tests, the settlement of the model foundation was measured at three corners by linear potentiometers. Based on the results of these tests, the following general conclusions can be drawn:

1. The settlement of the foundation under transient loading is generally uniform.
2. Failure in soil below the foundation may be in punching mode.

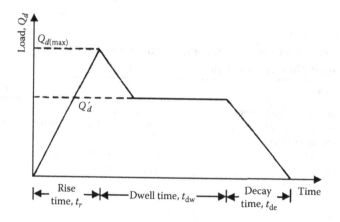

FIGURE 6.27 Nature of transient load in the laboratory tests of Cunny and Sloan. *Source:* Cunny, R. W., and R. C. Sloan. 1961. *Dynamic loading machine and results of preliminary small-scale footing tests*. Spec. Tech. Pub. 305, ASTM: 65.

3. Settlement under transient loading may be substantially less than that observed under static loading. As an example, for test 4 in Table 6.2, the settlement at ultimate load Q_u (static bearing capacity test) was about 66.55 mm. However, when subjected to a transient load with $Q_{d(max)} = 1.35$

TABLE 6.2
Load-Settlement Relationship of Square Surface Model Foundation on Sand Due to Transient Loading

Parameter	Test 1	Test 2	Test 3	Test 4
Width of model foundation B (mm)	152	203	203	229
Ultimate static load-carrying capacity Q_u (kN)	3.42	8.1	8.1	11.52
$Q_{d(max)}$ (kN)	3.56	13.97	10.12	15.57
Q'_d (kN)	3.56	12.45	9.67	14.46
$Q_{d(max)}/Q_u$	1.04	1.73	1.25	1.35
t_r (ms)	18	8	90	11
t_{dw} (ms)	122	420	280	0
t_{de} (ms)	110	255	290	350
S_e (Pot. 1) (mm)	7.11	—	21.08	10.16
S_e (Pot. 2) (mm)	1.27	—	23.62	10.67
S_e (Pot. 3) (mm)	2.79	—	24.13	10.16
Average S_e (mm)	3.73	—	22.94	10.34

Source: Compiled from Cunny, R. W., and R. C. Sloan. 1961. *Dynamic loading machine and results of preliminary small-scale footing tests*. Spec. Tech. Pub. 305, ASTM: 65.

TABLE 6.3

Load-Settlement Relationship of Square Surface Model Foundation on Clay Due to Transient Loading

Parameter	Test 1	Test 2	Test 3	Test 4
Width of model foundation B (mm)	114	114	114	127
Ultimate static load-carrying capacity Q_u (kN)	10.94	10.94	10.94	13.52
$Q_{d(max)}$ (kN)	12.68	13.79	15.39	15.92
Q'_d (kN)	10.12	12.54	13.21	13.12
$Q_{d(max)}/Q_u$	1.16	1.26	1.41	1.18
t_r (ms)	9	9	10	9
t_{dw} (ms)	170	0	0	0
t_{de} (ms)	350	380	365	360
S_e (Pot. 1) (mm)	12.7	16.76	43.18	14.73
S_e (Pot. 2) (mm)	12.7	18.29	42.67	13.97
S_e (Pot. 3) (mm)	12.19	17.78	43.18	13.97
Average S_e (mm)	12.52	17.60	43.00	14.22

Source: Compiled from Cunny, R. W., and R. C. Sloan. 1961. *Dynamic loading machine and results of preliminary small-scale footing tests.* Spec. Tech. Pub. 305, ASTM: 65.

Q_u, the observed settlement was about 10.4 mm. Similarly, for test 2 in Table 6.3, the settlement at ultimate load was about 51 mm. Under transient load with $Q_{d(max)} = 1.26\ Q_u$, the observed settlement was only about 18 mm.

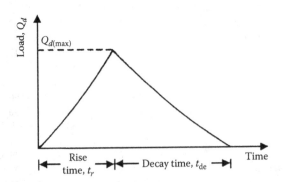

FIGURE 6.28 Nature of transient load in the laboratory tests of Jackson and Hadala. *Source:* Jackson, J. G., Jr., and P. F. Hadala. 1964. *Dynamic bearing capacity of soils. Report 3: The application similitude to small-scale footing tests.* U.S. Army Corps of Engineers, Waterways Experiment Station, Mississippi.

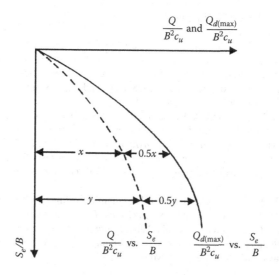

FIGURE 6.29 Relationship of $Q_{d(\max)}/B^2c_u$ versus S_e/B from plate load tests (plate size $B \times B$).

Jackson and Hadala[11] reported several laboratory model test results on square surface foundations with width B varying from 114 mm to 203 mm that were supported by saturated buckshot clay. For these tests, the nature of the transient load applied to the foundation is shown in Figure 6.28. The rise time t_r varied from 2 to 16 ms and the decay time from 240 to 425 ms. Based on these tests, it was shown that there is a unique relationship between $Q_{d(\max)}/(B^2c_u)$ and S_e/B. This relationship can be found in the following manner:

1. From the plate load test (square plate, $B \times B$) in the field, determine the relationship between load Q and S_e/B.
2. Plot a graph of Q/B^2c_u versus S_e/B as shown by the dashed line in Figure 6.29.
3. Since the strain-rate factor in clays is about 1.5 (see section 6.2), determine $1.5\ Q/B^2c_u$ and develop a plot of $1.5\ Q/B^2c_u$ versus S_e/B as shown by the solid line in Figure 6.29. This will be the relationship between $Q_{d(\max)}/(B^2c_u)$ versus S_e/B.

REFERENCES

1. Vesic, A. S., D. C. Banks, and J. M. Woodward. 1965. An experimental study of dynamic bearing capacity of footings on sand, in *Proceedings, VI Int. Conf. Soil Mech. Found. Eng.*, Montreal, Canada, 2: 209.
2. Vesic, A. S. 1973. Analysis of ultimate loads of shallow foundations. *J. Soil Mech. Found. Eng. Div.*, ASCE, 99(1): 45.
3. Whitman, R. V., and K. A. Healy. 1962. Shear strength of sands during rapid loading. *J. Soil Mech. Found. Eng. Div.*, ASCE, 88(2):99.

4. Carroll, W. F. 1963. *Dynamic bearing capacity of soils: Vertical displacement of spread footing on clay: Static and impulsive loadings*, Technical Report 3-599, Report 5, U.S. Army Corps of Engineers, Waterways Experiment Station, Mississippi.

5. Richards, R., Jr., D. G. Elms, and M. Budhu. 1993. Seismic bearing capacity and settlement of foundations. *J. Geotech. Eng.*, ASCE, 119(4): 622.

6. Raymond, G. P., and F. E. Komos. 1978. Repeated load testing of a model plane strain footing. *Canadian Geotech. J.* 15(2): 190.

7. Brumund, W. F., and G. A. Leonards. 1972. Subsidence of sand due to surface vibration. *J. Soil Mech. Found. Eng. Div.*, ASCE, 98(1): 27.

8. Das, B. M. 1993. *Principles of soil dynamics*. Boston, MA: PWS Publishers.

9. Das, B. M., and E. C. Shin. 1996. Laboratory model tests for cyclic load-induced settlement of a strip foundation on a clayey soil. *Geotech. Geol. Eng.*, London, 14: 213.

10. Cunny, R. W., and R. C. Sloan. 1961. Dynamic loading machine and results of preliminary small-scale footing tests. *Spec. Tech. Pub. 305*, ASTM: 65.

11. Jackson, J. G., Jr., and P. F. Hadala. 1964. *Dynamic bearing capacity of soils. Report 3: The application similitude to small-scale footing tests*. U.S. Army Corps of Engineers, Waterways Experiment Station, Mississippi.

7 Shallow Foundations on Reinforced Soil

7.1 INTRODUCTION

Reinforced soil, or mechanically stabilized soil, is a construction material that consists of soil that has been strengthened by tensile elements such as metal strips, geotextiles, or geogrids. In the 1960s, the French Road Research Laboratory conducted extensive research to evaluate the beneficial effects of using reinforced soil as a construction material. Results of the early work were well documented by Vidal.[1] During the last 40 years, many retaining walls and embankments were constructed all over the world using reinforced soil and they have performed very well.

The beneficial effects of soil reinforcement derive from (a) the soil's increased tensile strength and (b) the shear resistance developed from the friction at the soil-reinforcement interfaces. This is comparable to the reinforcement of concrete structures. At this time the design of reinforced earth is done with *free-draining granular soil only*. Thus, one avoids the effect of pore water pressure development in cohesive soil, which in turn controls the cohesive bond at the soil-reinforcement interfaces.

Since the mid-1970s a number of studies have been conducted to evaluate the possibility of constructing shallow foundations on reinforced soil to increase their load-bearing capacity and reduce settlement. In these studies, *metallic strips* and *geogrids* were used primarily as reinforcing material in *granular soil*. In the following sections the findings of these studies are summarized.

7.2 FOUNDATIONS ON METALLIC-STRIP–REINFORCED GRANULAR SOIL

7.2.1 METALLIC STRIPS

The metallic strips used for reinforcing granular soil for foundation construction are usually thin galvanized steel strips. These strips are laid in several layers under the foundation. For any given layer, the strips are laid at a given center-to-center spacing. The galvanized steel strips are subject to corrosion at the rate of about 0.025 to 0.05 mm per year. Hence, depending on the projected service life of a given structure, allowances must be made for the rate of corrosion during the design process.

7.2.2 FAILURE MODE

Binquet and Lee[2,3] conducted several laboratory tests and proposed a theory for designing a continuous foundation on sand reinforced with metallic strips. Figure 7.1 defines the general parameters in this design procedure. In Figure 7.1 the width of the continuous foundation is *B*. The first layer of reinforcement is placed at a distance

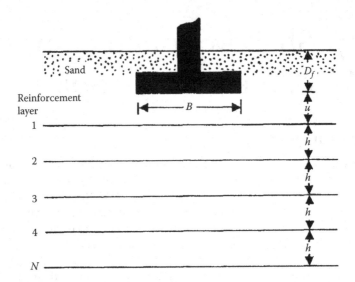

FIGURE 7.1 Foundation on metallic-strip-reinforced granular soil.

u measured from the bottom of the foundation. The distance between each layer of reinforcement is h. It was experimentally shown[2,3] that the most beneficial effect of reinforced earth is obtained when u/B is less than about two-thirds B and the number of layers of reinforcement N is greater than four but no more than six to seven. If the length of the ties (that is, reinforcement strips) is sufficiently long, failure occurs when the upper ties break. This phenomenon is shown in Figure 7.2.

Figure 7.3 shows an idealized condition for the development of a failure surface in reinforced earth that consists of two zones. Zone I is immediately below the foundation, which settles with the foundation during the application of load. In zone II the soil is pushed outward and upward. Points $A_1, A_2, A_3, \ldots$, and $B_1, B_2, B_3, \ldots$, which define the limits of zones I and II, are points at which maximum shear stress τ_{max} occurs in the xz plane. The distance $x = x'$ of the points measured from the center line of the foundation where maximum shear stress occurs is a function of z/B. This is shown in a nondimensional form in Figure 7.4.

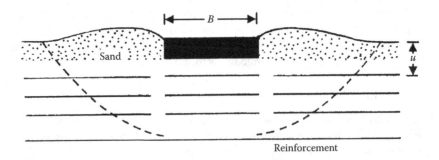

FIGURE 7.2 Failure in reinforced earth by tie break ($u/B < 2/3$ and $N \geq 4$).

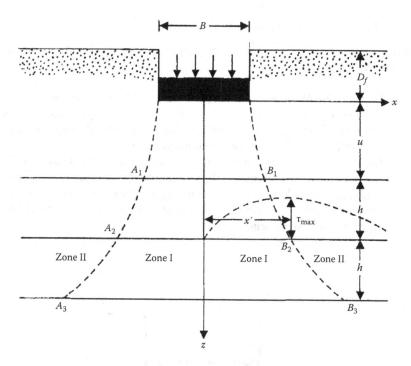

FIGURE 7.3 Failure surface in reinforced soil at ultimate load.

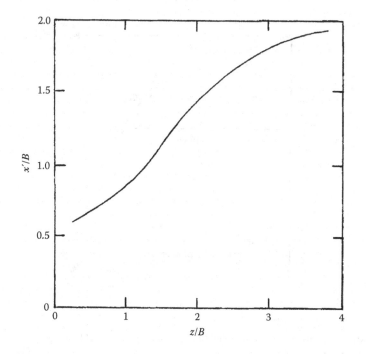

FIGURE 7.4 Variation of x'/B with z/B.

7.2.3   Forces in Reinforcement Ties

In order to obtain the forces in the reinforcement ties, Binquet and Lee[3] made the following assumptions:

1. Under the application of bearing pressure by the foundation, the reinforcing ties at points A_1, A_2, A_3, ... , and B_1, B_2, B_3, ... (Figure 7.3) take the shape shown in Figure 7.5a; that is, the tie takes two right angle turns on each side of zone I around two frictionless rollers.
2. For N reinforcing layers, the ratio of the load per unit area on the foundation supported by reinforced earth q_R to the load per unit area on the foundation supported by unreinforced earth q_o is constant, irrespective of the settlement level S_e (see Figure 7.5b). Binquet and Lee[2] proved this relation by laboratory experimental results.

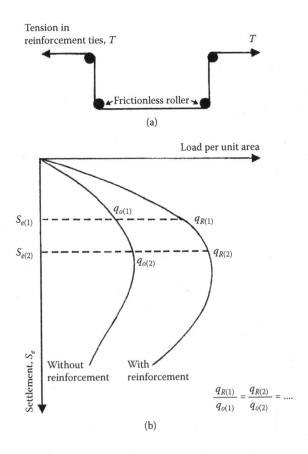

FIGURE 7.5 Assumptions to calculate the force in reinforcement ties.

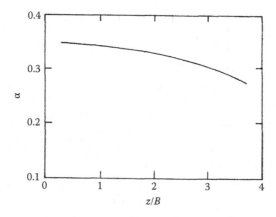

FIGURE 7.6 Variation of α with z/B.

With the above assumptions, it can be seen that

$$T = \frac{1}{N}\left[q_o\left(\frac{q_R}{q_o}-1\right)(\alpha B - \beta h)\right] \qquad (7.1)$$

where

T = tie force per unit length of the foundation at a depth z (kN/m)

N = number of reinforcement layers

q_o = load per unit area of the foundation on unreinforced soil for a foundation settlement level of $S_e = S'_e$

q_R = load per unit area of the foundation on reinforced soil for a foundation settlement level of $S_e = S'_e$

α, β = parameters that are functions of z/B

The variations of α and β with z/B are shown in Figures 7.6 and 7.7, respectively.

7.2.4 FACTOR OF SAFETY AGAINST TIE BREAKING AND TIE PULLOUT

In designing a foundation, it is essential to determine if the reinforcement ties will fail either by breaking or by pullout. Let the width of a single tie (at right angles to the cross section shown in Figure 7.1) be w and its thickness t. If the number of ties per unit length of the foundation placed at any depth z is equal to n, then the factor of safety against the possibility of tie break FS_B is

$$FS_B = \frac{wtnf_y}{T} = \frac{tf_y(\text{LDR})}{T} \qquad (7.2)$$

where

f_y = yield or breaking strength of tie material

LDR = linear density ratio = wn $\qquad (7.3)$

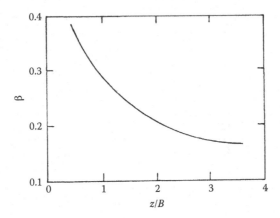

FIGURE 7.7 Variation of β with z/B.

Figure 7.8 shows a layer of reinforcement located at a depth z. The frictional resistance against tie pullout at that depth can be calculated as

$$F_p = 2\tan\phi_\mu \left[wn \int_{x=x'}^{x=X} \sigma\,dx + wn\gamma(X - x')(z + D_f) \right] \qquad (7.4)$$

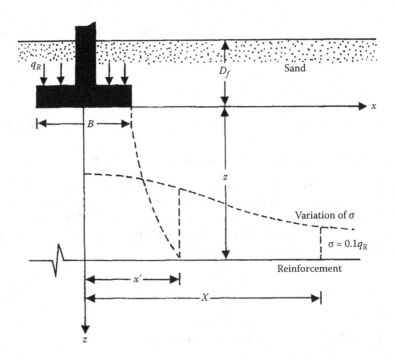

FIGURE 7.8 Frictional resistance against tie pullout.

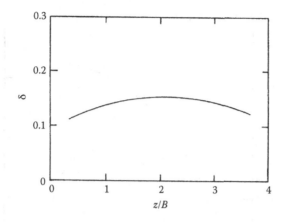

FIGURE 7.9 Variation of δ with z/B.

where

> ϕ_μ = soil–tie interface friction angle
> σ = effective normal stress at a depth z due to the uniform load per unit area q_R on the foundation
> X = distance at which $\sigma = 0.1q_R$
> D_f = depth of the foundation
> γ = unit weight of soil

Note that the second term in the right-hand side of equation (7.4) is due to the fact that frictional resistance is derived from the tops and bottoms of the ties. Thus, from equation (7.4),

$$F_p = 2\tan\phi_\mu(\text{LDR})\left[\delta Bq_o\left(\frac{q_R}{q_o}\right) + \gamma(X - x')(z + D_f)\right] \quad (7.5)$$

The term δ is a function of z/B and is shown in Figure 7.9. Figure 7.10 shows a plot of X/B versus z/B. Hence, at any given depth z, the factor of safety against tie pullout FS_P can be given as

$$FS_P = \frac{F_P}{T} \quad (7.6)$$

7.2.5 DESIGN PROCEDURE FOR A CONTINUOUS FOUNDATION

Following is a step-by-step procedure for designing a continuous foundation on granular soil reinforced with metallic strips.

Step 1. Establish the following parameters:
   A. Foundation:
      Net load per unit length Q
      Depth D_f

where

q_u = ultimate bearing capacity on unreinforced soil

$q = \gamma D_f$

N_q, N_γ = bearing capacity factors (Table 2.3)

Step 5. Determine the allowable bearing capacity q''_{all} based on allowable settlement as follows:

$$S_{e(\text{rigid})} = q''_{all} B \frac{(1-v^2)}{E_s} I$$

or

$$q''_{all} = \frac{E_s S_{e(\text{rigid})}}{B(1-v^2)I} \tag{7.9}$$

The variation of I with L/B (L = length of foundation) is given in Table 7.1.

Step 6. The smaller of the two allowable bearing capacities (that is, q'_{all} or q''_{all}) is equal to q_o.

Step 7. Calculate q_R (load per unit area of the foundation on reinforced soil) as

$$q_R = \frac{Q}{B} \tag{7.10}$$

Step 8. Calculate T for all layers of reinforcement using equation (7.1).

Step 9. Calculate the magnitude of F_p/T for each layer to see if $F_p/T \geq FS_p$. If $F_p/T < FS_p$, the length of the reinforcing strips may have to be increased by substituting X' ($>X$) in equation (7.5) so that F_p/T is equal to FS_p.

Step 10. Use equation (7.2) to obtain the thickness of the reinforcement strips.

Step 11. If the design is unsatisfactory, repeat steps 2 through 10.

TABLE 7.1
Variation of I with L/B

L/B	I
1	0.886
2	1.21
3	1.409
4	1.552
5	1.663
6	1.754
7	1.831
8	1.898
9	1.957
10	2.010

EXAMPLE 7.1

Design a continuous foundation with the following:

Foundation:
Net load to be carried $Q = 1.5$ MN/m
$D_f = 1.2$ m
Factor of safety against bearing capacity failure in unreinforced soil $F_s = 3.5$
Tolerable settlement $S_e = 25$ mm
Soil:
Unit weight $\gamma = 16.5$ kN/m³
Friction angle $\phi = 36°$
$E_s = 3.4 \times 10^4$ kN/m²
$v = 0.3$
Reinforcement ties:
Width $w = 70$ mm
$\phi_\mu = 25°$
$FS_B = 3$
$FS_P = 2$
$f_y = 2.5 \times 10^5$ kN/m²

Solution

Let $B = 1.2$ m, $u = 0.5$ m, $h = 0.5$ m, $N = 4$, and LDR = 60%. With LDR = 60%,

$$\text{Number of strips } n = \frac{LDR}{w} = \frac{0.6}{0.07} = 8.57/m$$

From equation (7.8),

$$q'_{all} = \frac{qN_q + \frac{1}{2}\gamma BN_\gamma}{FS}$$

From Table 2.3 for $\phi = 36°$, the magnitudes of N_q and N_γ are 37.75 and 44.43, respectively. So,

$$q'_{all} = \frac{(1.2 \times 16.5)(37.75) + (0.5)(16.5)(1.2)(44.43)}{3.5} = 339.23 \text{ kN/m}^2$$

From equation (7.9),

$$q''_{all} = \frac{E_s S_e}{B(1-v^2)I} = \frac{(3.4 \times 10^4)(0.025)}{(1.2)[1-(0.3)^2](2)} = 389.2 \text{ kN/m}^2$$

Since $q''_{all} > q'_{all}$, $q_o = q'_{all} = 339.23$ kN/m². Thus,

$$q_R = \frac{Q}{B} = \frac{1.5 \times 10^3 \text{ kN}}{1.2} = 1250 \text{ kN/m}^2$$

Now the tie forces can be calculated using equation (7.1):

$$T = \left(\frac{q_o}{N}\right)\left(\frac{q_R}{q_o} - 1\right)(\alpha B - \beta h)$$

Layer No.	$\left(\dfrac{q_o}{N}\right)\left(\dfrac{q_R}{q_o}-1\right)$	z (m)	z/B	$\alpha B - \beta h$	T (kN/m)
1	227.7	0.5	0.47	0.285	64.89
2	227.7	1.0	0.83	0.300	68.31
3	227.7	1.5	1.25	0.325	74.00
4	227.7	2.0	1.67	0.330	75.14

Note: B = 1.2 m; α from Figure 7.6; β from Figure 7.7; *h* = 0.5 m.

The magnitudes of F_p/T for each layer are calculated in the following table. From equations (7.5) and (7.6),

$$\frac{F_p}{T} = \frac{2\tan\phi_\mu(LDR)}{T}\left[\delta Bq_o\left(\frac{q_R}{q_o}\right)+\gamma(X-x')(z+D_f)\right]$$

Parameter	Layer			
	1	2	3	4
$\dfrac{2\tan\phi_\mu(LDR)}{T}$ (m/kN)	0.0086	0.0082	0.0076	0.0075
z/B	0.47	0.83	1.25	1.67
δ	0.12	0.14	0.15	0.16
$\delta Bq_o\left(\dfrac{q_R}{q_o}\right)$ (kN/m)	180	210	225	240
X/B	1.4	2.3	3.2	3.6
X (m)	1.68	2.76	3.84	4.32
x'/B	0.7	0.8	1.0	1.3
x' (m)	0.84	0.96	1.2	1.56
$\gamma(X-x')(z+D_f)$ (kN/m)	23.56	65.34	117.6	145.7
F_p/T	1.75	2.26	2.6	2.89

The minimum factor of safety FS_p required is two. In all layers except layer 1, F_p/T is greater than two. So we need to find a new value of $x = X'$ so that F_p/T is equal to two. So, for layer 1,

$$\frac{F_p}{T} = \frac{2\tan\phi_\mu(LDR)}{T}\left[\delta Bq_o\left(\frac{q_R}{q_o}\right)+\gamma(X-x')(z+D_f)\right]$$

or

$$2 = 0.0086[180+16.5(X'-0.84)(0.5+1.2)]; \ X' = 2.71 \text{ m}$$

Tie thickness *t*:

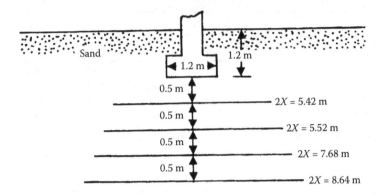

FIGURE 7.11 Length of reinforcement under the foundation.

From equation (7.2),

$$FS_B = \frac{tf_y(\text{LDR})}{T}$$

$$t = \frac{(FS_B)(T)}{(f_y)(\text{LDR})} = \frac{(3)(T)}{(2.5 \times 10^5)(0.6)} = 2 \times 10^{-5} T$$

The following table can now be prepared:

Layer No.	T (kN/m)	t (mm)
1	64.89	≈1.3
2	68.31	≈1.4
3	74.00	≈1.5
4	75.14	≈1.503

A tie thickness of **1.6 mm** will be sufficient for all layers. Figure 7.11 shows a diagram of the foundation with the ties.

7.3 FOUNDATIONS ON GEOGRID-REINFORCED GRANULAR SOIL

7.3.1 GEOGRIDS

A geogrid is defined as a polymeric (that is, geosynthetic) material consisting of connected parallel sets of tensile ribs with apertures of sufficient size to allow strike-through of surrounding soil, stone, or other geotechnical material. Its primary function is reinforcement. Reinforcement refers to the mechanism(s) by which the engineering properties of the composite soil/aggregate can be mechanically improved.

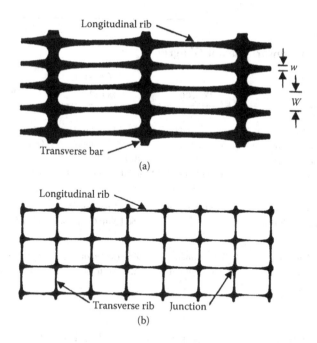

FIGURE 7.12 Geogrids: (a) uniaxial; (b) biaxial.

Commercially available geogrids may be categorized by manufacturing process, principally (a) extruded, (b) woven, and (c) welded. Extruded geogrids are formed using a thick sheet of polyethylene or polypropylene that is punched and drawn to create apertures and to enhance the engineering properties of the resulting ribs and nodes. Woven geogrids are manufactured by grouping polymerics—usually polyester or polypropylene—and weaving them into a mesh pattern that is then coated with a polymeric lacquer. Welded geogrids are manufactured by fusing junctions of polymeric strips. Extruded geogrids have shown good performance when compared to other types for pavement reinforcement applications.

Geogrids generally are of two types: (a) biaxial geogrids and (b) uniaxial geogrids. Figure 7.12 shows the two types of described geogrids that are produced by Tensar International. Uniaxial Tensar® grids are manufactured by stretching a punched sheet of extruded high-density polyethylene in one direction under carefully controlled conditions. This process aligns the polymer's long-chain molecules in the direction of draw and results in a product with high one-directional tensile strength and modulus. Biaxial Tensar® grids are manufactured by stretching the punched sheet of polypropylene in two orthogonal directions. This process results in a product with high tensile strength and modulus in two perpendicular directions. The resulting grid apertures are either square or rectangular.

As mentioned previously, there are several types of geogrids that are commercially available in different countries now. The commercial geogrids currently available for soil reinforcement have nominal rib thicknesses of about 0.5–1.5 mm and junctions of about 2.5–5 mm. The grids used for soil reinforcement usually have

TABLE 7.2
Properties of Tensar® Biaxial Geogrids

Property	Geogrid		
	BX1100	BX1200	BX1300
Aperture size			
Machine direction	25 mm (nominal)	25 mm (nominal)	46 mm (nominal)
Cross-machine direction	33 mm (nominal)	33 mm (nominal)	64 mm (nominal)
Open area	70% (minimum)	70% (minimum)	75% (minimum)
Junction thickness	2.9 mm (nominal)	4.0 mm (nominal)	4.4 mm (nominal)
Tensile modulus			
Machine direction	205 kN/m (nominal)	300 kN/m (nominal)	275 kN/m (nominal)
Cross-machine direction	330 kN/m (nominal)	450 kN/m (nominal)	475 kN/m (nominal)
Material			
Polypropylene	99% (minimum)	99% (minimum)	98% (minimum)
Carbon black	0.5% (minimum)	0.5% (minimum)	1.3% (minimum)

apertures that are rectangular or elliptical in shape. The dimensions of the apertures vary from about 25 to 160 mm. Geogrids are manufactured so that the open areas of the grids are greater than 50% of the total area. They develop reinforcing strength at low strain levels (such as 2%). Table 7.2 gives some properties of the Tensar® biaxial geogrids that are currently available commercially.

7.3.2 GENERAL PARAMETERS

Since the mid-1980s, a number of laboratory model studies have been reported relating to the evaluation of the ultimate and allowable bearing capacities of shallow foundations supported by soil reinforced with multiple layers of geogrids. The results obtained so far seem promising. The general parameters of the problem are defined in this section.

Figure 7.13 shows the general parameters of a rectangular surface foundation on a soil layer reinforced with several layers of geogrids. The size of the foundation is $B \times L$ (width × length) and the size of the geogrid layers is $b \times l$ (width × length). The first layer of geogrid is located at a depth u below the foundation, and the vertical distance between consecutive layers of geogrids is h. The total depth of reinforcement d can be given as

$$d = u + (N - 1)h \tag{7.11}$$

where
   N = number of reinforcement layers

The beneficial effects of reinforcement to increase the bearing capacity can be expressed in terms of a nondimensional parameter called the bearing capacity ratio

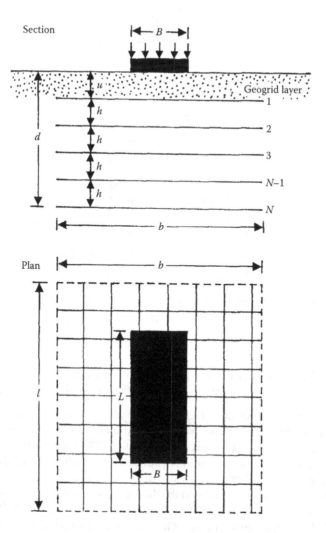

FIGURE 7.13 Geometric parameters of a rectangular foundation supported by geogrid-reinforced soil.

(BCR). The bearing capacity ratio can be expressed with respect to the ultimate bearing capacity or the allowable bearing capacity (at a given settlement level of the foundation). Figure 7.14 shows the general nature of the load-settlement curve of a foundation both with and without geogrid reinforcement. Based on this concept the bearing capacity ratio can be defined as

$$\text{BCR}_u = \frac{q_{u(R)}}{q_u} \tag{7.12}$$

and

$$\text{BCR}_s = \frac{q_R}{q} \tag{7.13}$$

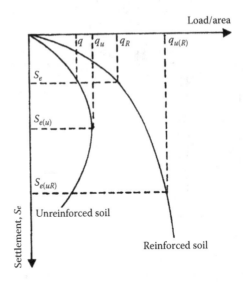

FIGURE 7.14 General nature of the load-settlement curves for unreinforced and geogrid-reinforced soil supporting a foundation.

where

    BCR_u = bearing capacity ratio with respect to the ultimate load

    BCR_s = bearing capacity ratio at a given settlement level S_e for the foundation

For a given foundation and given values of b/B, l/B, u/B, and h/B, the magnitude of BCR_u increases with d/B and reaches a maximum value at $(d/B)_{cr}$, beyond which the bearing capacity remains practically constant. The term $(d/B)_{cr}$ is the critical-reinforcement-depth ratio. For given values of l/B, u/B, h/B, and d/B, BCR_u attains a maximum value at $(b/B)_{cr}$, which is called the critical-width ratio. Similarly, a critical-length ratio $(l/B)_{cr}$ can be established (for given values of b/B, u/B, h/B, and d/B) for a maximum value of BCR_u. This concept is schematically illustrated in Figure 7.15. As an example, Figure 7.16 shows the variation of BCR_u with d/B for four model foundations ($B/L = 0$, 1/3, 1/2, and 1) as reported by Omar et al.[4] It was also shown from laboratory model tests[4,5] that, for a given foundation, if b/B, l/B, d/B, and h/B are kept constant, the nature of variation of BCR_u with u/B will be as shown in Figure 7.17. Initially (zone 1), BCR_u increases with u/B to a maximum value at $(u/B)_{cr}$. For $u/B > (u/B)_{cr}$, the magnitude of BCR_u decreases (zone 2). For $u/B > (u/B)_{max}$, the plot of BCR_u versus u/B generally flattens out (zone 3).

7.3.3   RELATIONSHIPS FOR CRITICAL NONDIMENSIONAL PARAMETERS FOR FOUNDATIONS ON GEOGRID-REINFORCED SAND

Based on the results of their model tests and other existing results, Omar et al.[4] developed the following empirical relationships for the nondimensional parameters $(d/B)_{cr}$, $(b/B)_{cr}$, and $(l/B)_{cr}$ described in the preceding section:

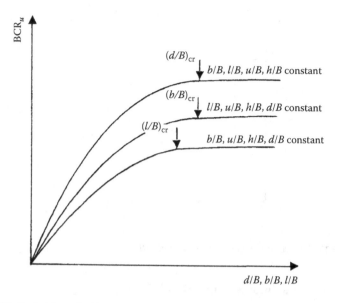

FIGURE 7.15 Definition of critical nondimensional parameters—$(d/B)_{cr}$, $(b/B)_{cr}$, and $(l/B)_{cr}$.

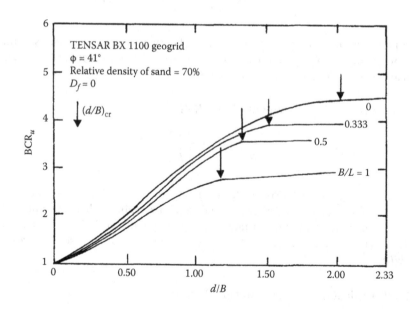

FIGURE 7.16 Variation of BCR_u with d/B. *Source:* Based on the results of Omar, M. T., B. M. Das, S. C. Yen, V. K. Puri, and E. E. Cook. 1993. Ultimate bearing capacity of rectangular foundations on geogrid-reinforced sand. *Geotech. Testing J.,* ASTM, 16(2): 246.

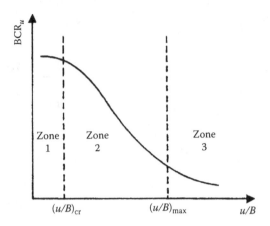

FIGURE 7.17 Nature of variation of BCR_u with u/B.

7.3.3.1 Critical Reinforcement—Depth Ratio

$$\left(\frac{d}{b}\right)_{cr} = 2 - 1.4\left(\frac{B}{L}\right) \quad \left(\text{for } 0 \le \frac{B}{L} \le 0.5\right) \tag{7.14}$$

$$\left(\frac{d}{b}\right)_{cr} = 1.43 - 0.26\left(\frac{B}{L}\right) \quad \left(\text{for } 0.5 \le \frac{B}{L} \le 1\right) \tag{7.15}$$

The preceding relationships suggest that the bearing capacity increase is realized only when the reinforcement is located within a depth of $2B$ for a continuous foundation and a depth of $1.2B$ for a square foundation.

7.3.3.2 Critical Reinforcement–Width Ratio

$$\left(\frac{b}{B}\right)_{cr} = 8 - 3.5\left(\frac{B}{L}\right)^{0.51} \tag{7.16}$$

According to equation (7.16), $(b/B)_{cr}$ is about 8 for a continuous foundation and about 4.5 for a square foundation. It needs to be realized that, generally, with other parameters remaining constant, about 80% or more of BCR_u is realized with $b/B \approx 2$. The remaining 20% of BCR_u is realized when b/B increases from about 2 to $(b/B)_{cr}$.

7.3.3.3 Critical Reinforcement–Length Ratio

$$\left(\frac{l}{B}\right)_{cr} = 3.5\left(\frac{B}{L}\right) + \frac{L}{B} \tag{7.17}$$

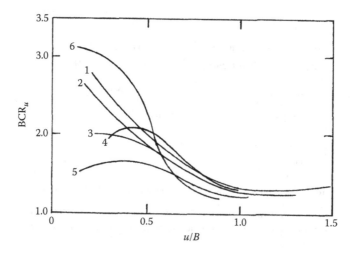

FIGURE 7.18 Variation of BCR_u with u/B from various published works (see Table 7.3 for details).

7.3.3.4 Critical Value of u/B

Figure 7.18 shows the laboratory model test results of Guido et al.,[5] Akinmusuru and Akinbolade[6] and Yetimoglu et al.[7] for bearing capacity tests conducted on *surface foundations* supported by multi-layered reinforced sand. Details of these tests are given in Table 7.3. Based on the definition given in Figure 7.17, it appears from these test results that $(u/B)_{max} \approx 0.9$ to 1. From Figure 7.18 it may also be seen that $(u/B)_{cr}$ as defined by Figure 7.17 is about 0.25 to 0.5. An analysis of the test results of Schlosser et al.[8] yields a value of $(u/B)_{cr} \approx 0.4$. Large-scale model tests by Adams and Collin[9] showed that $(u/B)_{cr}$ is approximately 0.25.

TABLE 7.3
Details of Test Parameters for Plots Shown in Figure 7.18

Curve	Investigator	Type of Model Foundation	Type of Reinforcement	Parametric Details
1	Guido et al.[5]	Square	Tensar® BX1100	$h/B = 0.25$;
2	Guido et al.[5]	Square	Tensar® BX1200	$b/B = 3$;
3	Guido et al.[5]	Square	Tensar® BX1300	$N = 3$
4	Akinmusuru and Akinbolade[6]	Square	Rope fibers	$h/B = 0.5$; $b/B = 3$; $N = 5$
5	Yetimoglu et al.[7]	Rectangular; $B/L = 8$; L = length of foundation	Terragrid® GS100	$b/B = 4$; $N = 1$
6	Yetimoglu et al.[7]	Rectangular; $B/L = 8$; L = length of foundation	Terragrid® GS100	$h/B = 0.3$; $b/B = 4.5$; $N = 4$

TABLE 7.4
Physical Properties of the Geogrid Used by Shin and Das for the Results Shown in Figure 7.19

Physical Property	Value
Polymer type	Polypropylene
Structure	Biaxial
Mass per unit area	320 g/m^2
Aperture size	41 mm (MD) × 31 mm (CMD)
Maximum tensile strength	14.5 kN/m (MD) × 20.5 kN/m (CMD)
Tensile strength at 5% strain	5.5 kN/m (MD) × 16.0 kN/m (CMD)

Source: Shin, E. C., and B. M. Das. 2000. Experimental study of bearing capacity of a strip foundation on geogrid-reinforced sand. *Geosynthetics Intl.* 7(1): 59.

Note: CMD, cross-machine direction; MD, machine direction.

7.3.4 BCR$_u$ for Foundations with Depth of Foundation D_f Greater Than Zero

To the best of the author's knowledge, the only tests for bearing capacity of shallow foundations with $D_f > 0$ are those reported by Shin and Das.[10] These results were for laboratory model tests on a strip foundation in sand. The physical properties of the geogrid used in these tests are given in Table 7.4.

The model tests were conducted with d/B from 0 to 2.4, $u/B = 0.4$, $h/B = 0.4$, and $b/B = 6$ [≈ $(b/B)_{cr}$]. The sand had relative densities D_r of 59% and 74%, and D_f/B was varied from 0 to 0.75. The variation of BCR$_u$ with d/B, D_f/B, and D_r is shown in Figure 7.19. From this figure the following observations can be made:

1. For all values of D_f/B and D_r, the magnitude of $(d/B)_{cr}$ is about two for strip foundations.
2. For given b/B, D_r, u/B, and h/B, the magnitude of BCR$_u$ increases with D_f/B.

Based on their laboratory model test results, Das and Shin[10] have shown that the ratio of BCR$_s$:BCR$_u$ for strip foundations has an approximate relationship with the embedment ratio (D_f:B) for a settlement ratio S_e:B less than or equal to 5%. This relationship is shown in Figure 7.20 and is valid for any values of d/B and b/B. The definition of BCR$_s$ was given in equation (7.13).

7.3.4.1 Settlement at Ultimate Load

As shown in Figure 7.14, a foundation supported by geogrid-reinforced sand shows a greater level of settlement at ultimate load $q_{u(R)}$. Huang and Hong[11] analyzed the laboratory test results of Huang and Tatsuoka,[12] Takemura et al.,[13] Khing et al.,[14] and

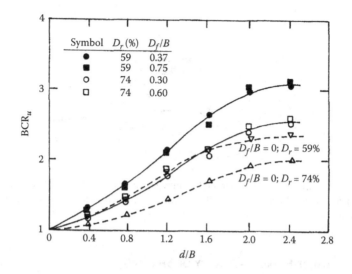

FIGURE 7.19 Comparison of BCR_u for tests conducted at $D_f/B = 0$ and $D_f/B > 0$—strip foundation; $u/B = h/B = 0.4$; $B = 67$ mm; $b/B = 6$. *Source:* Compiled from the results of Shin, E. C., and B. M. Das. 2000. Experimental study of bearing capacity of a strip foundation on geogrid-reinforced sand. *Geosynthetics Intl.* 7(1): 59.

Yetimoglu et al.[7] and provided the following approximate relationship for settlement at ultimate load. Or,

$$\frac{S_{e(uR)}}{S_{e(u)}} = 1 + 0.385(BCR_u - 1) \tag{7.18}$$

Refer to Figure 7.14 for definitions of $S_{e(uR)}$ and $S_{e(u)}$.

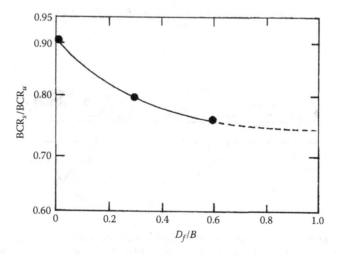

FIGURE 7.20 Plot of BCR_s/BCR_u with D_f/B (at settlement ratios < 5%).

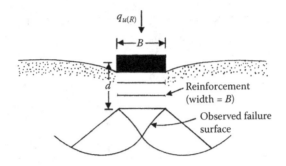

FIGURE 7.21 Failure surface observed by Huang and Tatsuoka. *Source:* From Huang, C. C., and F. Tatsuoka. 1990. Bearing capacity of reinforced horizontal sandy ground. *Geotextiles and Geomembranes.* 9: 51.

7.3.5 ULTIMATE BEARING CAPACITY OF SHALLOW FOUNDATIONS ON GEOGRID-REINFORCED SAND

Huang and Tatsuoka[12] proposed a failure mechanism for a strip foundation supported by reinforced earth where the width of reinforcement b is equal to the width of the foundation B, and this is shown in Figure 7.21. This is the so-called *deep foundation mechanism* where a quasi-rigid zone is developed beneath the foundation. Schlosser[8] proposed a *wide slab mechanism* of failure in soil at ultimate load for the condition where $b > B$, and this is shown in Figure 7.22. Huang and Meng[15] provided an analysis to estimate the ultimate bearing capacity of *surface* foundations supported by geogrid-reinforced sand. This analysis took into account the wide slab mechanism as shown in Figure 7.22. According to this analysis and referring to Figure 7.22,

$$q_{u(R)} = \left[0.5 - 0.1 \left(\frac{B}{L} \right) \right] (B + \Delta B) \gamma B N_\gamma + \gamma d N_q \qquad (7.19)$$

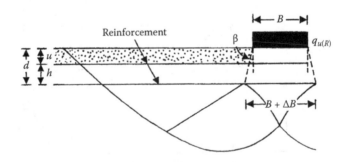

FIGURE 7.22 Failure mechanism of reinforced ground proposed by Schlosser et al. *Source:* From Schlosser, F., H. M. Jacobsen, and I. Juran. 1983. Soil reinforcement—general report. *Proc. VIII European Conf. Soil Mech. Found. Engg.* Helsinki, Balkema, 83.

where
    L = length of foundation
    γ = unit weight of soil
and
    $\Delta B = 2d \tan \beta$

$$(7.20)$$

The relationships for the bearing capacity factors N_γ and N_q are given in equations (2.66) and (2.74) (see Table 2.3 for values of N_q and Table 2.4 for values of N_γ).
    The angle β is given by the relation

$$\tan \beta = 0.68 - 2.071\left(\frac{h}{B}\right) + 0.743(\text{CR}) + 0.03\left(\frac{b}{B}\right) \qquad (7.21)$$

where

$$\text{CR} = \text{cover ratio} = \frac{\text{width of reinforcing strip}}{\text{center-to-center horizontal spacing of the strips}} = \frac{w}{W} \quad \text{(Figure 7.12)}$$

Equation (7.21) is valid for the following ranges:

$$0 \le \tan \beta \le 1 \qquad 1 \le \frac{b}{B} \le 10$$

$$0.25 \le \frac{h}{B} \le 0.5 \qquad 1 \le N \le 5$$

$$0.02 \le \text{CR} \le 1.0 \qquad 0.3 \le \frac{d}{B} \le 2.5$$

In equation (7.21), it is important to note that the parameter h/B plays the *primary* role in predicting β, and CR plays the secondary role. The effect of b/B is small.

7.3.6 TENTATIVE GUIDELINES FOR BEARING CAPACITY CALCULATION IN SAND

Considering the bearing capacity theories presented in the preceding section, following is a tentative guideline (mostly conservative) for estimating the ultimate and allowable bearing capacities of foundations supported by geogrid-reinforced sand:

    Step 1. The magnitude of u/B should be kept between 0.25 and 0.33.
    Step 2. The value of h/B should not exceed 0.4.
    Step 3. For most practical purposes and for economic efficiency, b/B should be kept between 2 and 3 and $N \le 4$.
    Step 4. Use equation (7.19), slightly modified, to calculate $q_{u(R)}$, or

$$q_{u(R)} = \left[0.5 - 0.1\left(\frac{B}{L}\right)\right](B + 2d \tan \beta)\gamma N_\gamma + \gamma(D_f + d)N_q \qquad (7.22)$$

where

$$\beta \approx \tan^{-1}\left[0.68-2.071\left(\frac{h}{B}\right)+0.743(\text{CR})+0.03\left(\frac{b}{B}\right)\right] \quad (7.23)$$

Step 5. For determining q_R at $S_e/B \le 5\%$,

a. Calculation of $\text{BCR}_u = q_{u(R)}/q_u$. The relationship for $q_{u(R)}$ is given in equation (7.22). Also,

$$q_u = \left[0.5-0.1\left(\frac{B}{L}\right)\right]B\gamma N_\gamma + \gamma D_f N_q \quad (7.24)$$

b. With known values of D_f/B and using Figure 7.20, obtain $\text{BCR}_s/\text{BCR}_u$.

c. From steps a and b, obtain $\text{BCR}_s = q_R/q$.

d. Estimate q from the relationships given in equations (5.43) and (5.44) as

$$q = \frac{S_e N_{60}}{1.25\left[1-\left(\frac{D_f}{4B}\right)\right]} = \frac{0.8 S_e N_{60}}{1-\left(\frac{D_f}{4B}\right)} \quad \text{(for } B \le 1.22 \text{ m)} \quad (7.25)$$

and

$$q = \frac{S_e N_{60}}{2\left[1-\left(\frac{D_f}{4B}\right)\right]}\left(\frac{B+0.3}{B}\right)^2 = \frac{0.5 S_e N_{60}}{1-\left(\frac{D_f}{4B}\right)}\left(\frac{B+0.3}{B}\right)^2 \quad \text{(for } B > 1.22 \text{ m)} \quad (7.26)$$

where
q is in kN/m², S_e is in mm, D_f and B are in m
N_{60} = average field standard penetration number

e. Calculate $q_R = q(\text{BCR}_s)$

7.3.7 BEARING CAPACITY OF ECCENTRICALLY LOADED STRIP FOUNDATION

Patra et al.[16] conducted several model tests in the laboratory to determine the ultimate bearing capacity of eccentrically loaded strip foundations. The width of the foundation B was 80 mm for these tests. For geogrid reinforcement, u/B, h/B, and b/B were kept equal to 0.35, 0.25, and 5, respectively. The relative density of sand was 72% for all tests. Based on these test results, it was proposed that

$$\frac{q_{u(R)-e}}{q_{u(R)}} = 1-R_{KR} \quad (7.27)$$

where
$q_{u(R)-e}$ and $q_{u(R)}$ = ultimate bearing capacity with load eccentricities $e > 0$ and $e = 0$, respectively
R_{KR} = reduction factor

The reduction factor is given by the relation

$$R_{KR} = 4.97 \left(\frac{D_f + d}{B} \right)^{-0.12} \left(\frac{e}{B} \right)^{1.21} \tag{7.28}$$

For these tests the depth of the foundation was varied from zero to $D_f = B$.

7.3.8 SETTLEMENT OF FOUNDATIONS ON GEOGRID-REINFORCED SOIL DUE TO CYCLIC LOADING

In many cases, shallow foundations supported by geogrid-reinforced soil may be subjected to cyclic loading. This problem will primarily be encountered by vibratory machine foundations. Das[17] reported laboratory model test results on settlement caused by cyclic loading on surface foundations supported by reinforced sand. The results of the tests are summarized below.

The model tests were conducted with a square model foundation on unreinforced and geogrid-reinforced sand. Details of the sand and geogrid parameters were:

Model foundation:
   Square; $B = 76.2$ mm
Sand:
   Relative density of compaction $D_r = 76\%$
   Angle of friction $\phi = 42°$
Reinforcement:
   Geogrid: Tensar® BX1000
Reinforcement-width ratio:

$$\left(\frac{b}{B} \right) \approx \left(\frac{b}{B} \right)_{cr} \text{ [see equation (7.16)]}$$

$$\left(\frac{u}{B} \right) \approx \left(\frac{u}{B} \right)_{cr} = 0.33$$

$$\frac{h}{B} = 0.33$$

Reinforcement-depth ratio:

$$\left(\frac{d}{B} \right) \approx \left(\frac{d}{B} \right)_{cr} = 1.33 \text{ [see equation (7.15)]}$$

Number of layers of reinforcement:
   $N = 4$

The laboratory tests were conducted by first applying a static load of intensity q_s ($= q_{u(R)}/FS$; FS = factor of safety) followed by a cyclic load of low frequency (1 cps).

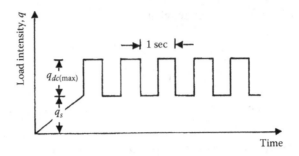

FIGURE 7.23 Nature of load application—cyclic load test.

The amplitude of the intensity of cyclic load was $q_{dc(max)}$. The nature of load application described is shown in Figure 7.23. Figure 7.24 shows the nature of variation of foundation settlement due to cyclic load application S_{ec} with $q_{dc(max)}/q_{u(R)}$ and number of load cycles n. This is for the case of $FS = 3$. Note that, for any given test, S_{ec} increases with n and reaches practically a maximum value $S_{ec(max)}$ at $n = n_{cr}$. Based on these tests the following conclusions can be drawn:

1. For given values of FS and n, the magnitude of S_{ec}/B increases with the increase in $q_{dc(max)}/q_{u(R)}$.

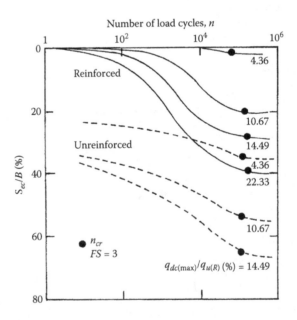

FIGURE 7.24 Plot of S_{ec}/B versus n. (*Note*: For reinforced sand, $u/B = h/B = 1/3$; $b/B = 4$; $d/B = 1$–1/3.) *Source:* After Das, B. M. 1998. Dynamic loading on foundation on reinforced soil, in *Geosynthetics in foundation reinforcement and erosion control systems*, eds. J. J. Bowders, H. B. Scranton, and G. P. Broderick. Geotech. Special Pub. 76, ASCE, 19.

2. If the magnitudes of $q_{dc(max)}/q_{u(R)}$ and n remain constant, the value of S_{ec}/B increases with a decrease in FS.
3. The magnitude of n_{cr} for all tests in reinforced soil is approximately the same, varying between 1.75×10^5 and 2.5×10^5 cycles. Similarly, the magnitude of n_{cr} for all tests in unreinforced soil varies between 1.5×10^5 and 2.0×10^5 cycles.

The variations of $S_{ec(max)}/B$ obtained from these tests for various values of $q_{dc(max)}/q_{u(R)}$ and FS are shown in Figure 7.25. This figure clearly demonstrates the reduction

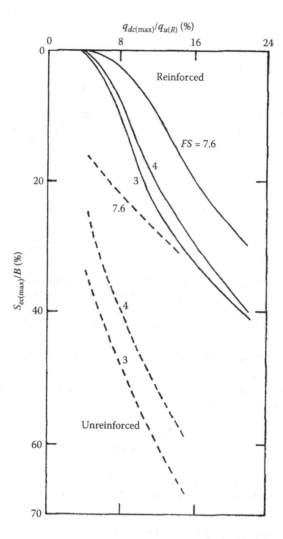

FIGURE 7.25 Plot of $S_{ec(max)}/B$ versus $q_{dc(max)}/q_{u(R)}$. (*Note:* For reinforced sand, $u/B = h/B = 1/3$; $b/B = 4$; $d/B = 1\text{--}1/3$.) *Source:* After Das, B. M. 1998. Dynamic loading on foundation on reinforced soil, in *Geosynthetics in foundation reinforcement and erosion control systems*, eds. J. J. Bowders, H. B. Scranton, and G. P. Broderick. Geotech. Special Pub. 76, ASCE, 19.

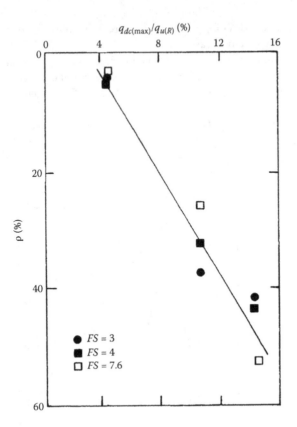

FIGURE 7.26 Variation of $q_{dc(max)}/q_{u(R)}$ with ρ. (*Note:* For reinforced sand, $u/B = h/B = 1/3$; $b/B = 4$; $d/B = 1\text{--}1/3$.) *Source:* After Das, B. M. 1998. Dynamic loading on foundation on reinforced soil, in *Geosynthetics in foundation reinforcement and erosion control systems*, eds. J. J. Bowders, H. B. Scranton, and G. P. Broderick. Geotech. Special Pub. 76, ASCE, 19.

of the level of permanent settlement caused by geogrid reinforcement due to cyclic loading. Using the results of $S_{ec(max)}$ given in Figure 7.25, the variation of settlement ratio ρ for various combinations of $q_{dc(max)}/q_{u(R)}$ and FS are plotted in Figure 7.26. The settlement ratio is defined as

$$\rho = \frac{S_{ec(max)} - \text{reinforced}}{S_{ec(max)} - \text{unreinforced}} \tag{7.29}$$

From Figure 7.26 it can be seen that, although some scattering exists, the settlement ratio is only a function of $q_{dc(max)}/q_{u(R)}$ and not the factor of safety FS.

7.3.9 Settlement Due to Impact Loading

Geogrid reinforcement can reduce the settlement of shallow foundations that are likely to be subjected to impact loading. This is shown in the results of laboratory

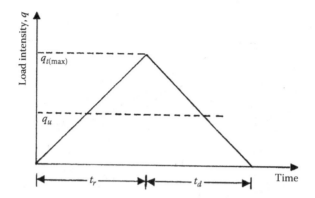

FIGURE 7.27 Nature of transient load.

model tests in sand reported by Das.[17] The tests were conducted with a square surface foundation ($D_f = 0$; $B = 76.2$ mm). Tensar® BX1000 geogrid was used as reinforcement. Following are the physical parameters of the soil and reinforcement:

Sand:
  Relative density of compaction = 76%
  Angle of friction $\phi = 42°$
Reinforcement:

$$\frac{u}{B} = 0.33; \quad \frac{b}{B} = 4; \quad \frac{h}{B} = 0.33$$

Number of reinforcement layers $N = 0, 1, 2, 3,$ and 4

The idealized shape of the impact load applied to the model foundation is shown in Figure 7.27, in which t_r and t_d are the rise and decay times and $q_{t(max)}$ is the maximum intensity of the impact load. For these tests the average values of t_r and t_d were approximately 1.75 s and 1.4 s, respectively. The maximum settlements observed due to the impact loading $S_{et(max)}$ are shown in a nondimensional form in Figure 7.28. In this figure q_u and $S_{e(u)}$, respectively, are the ultimate bearing capacity and the corresponding foundation settlement on unreinforced sand. From this figure it is obvious that

1. For a given value of $q_{t(max)}/q_u$, the foundation settlement decreases with an increase in the number of geogrid layers.
2. For a given number of reinforcement layers, the magnitude of $S_{et(max)}$ increases with the increase in $q_{t(max)}/q_u$.

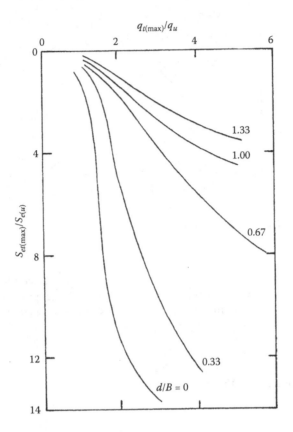

FIGURE 7.28 Variation of $S_{et\,(max)}/S_{e(u)}$ with $q_{t\,(max)}/q_u$ and $\frac{d}{B}$ d/B. *Source: After Das, B. M. 1998. Dynamic loading on foundation on reinforced soil, in Geosynthetics in foundation reinforcement and erosion control systems, eds. J. J. Bowders, H. B. Scranton, and G. P. Broderick. Geotech. Special Pub. 76, ASCE: 19.*

The effectiveness with which geogrid reinforcement helps reduce the settlement can be expressed by a quantity called the settlement reduction factor R, or

$$R = \frac{S_{et(max)-d}}{S_{et(max)-d=0}}$$

where

$S_{et(max)-d}$ = maximum settlement due to impact load with reinforcement depth of d

$S_{et(max)-d=0}$ = maximum settlement with no reinforcement (that is, $d = 0$ or $N = 0$)

Based on the results given in Figure 7.28, the variation of R with $q_{t(max)}/q_u$ and d/d_{cr} is shown in Figure 7.29. From the plot it is obvious that the geogrid reinforcement acts as an excellent settlement retardant under impact loading.

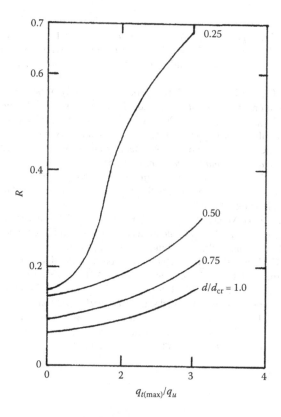

FIGURE 7.29 Plot of settlement reduction factor with $q_{t(\max)}/q_u$ and d/d_{cr}. *Source:* After Das, B. M. 1998. Dynamic loading on foundation on reinforced soil, in *Geosynthetics in foundation reinforcement and erosion control systems*, eds. J. J. Bowders, H. B. Scranton, and G. P. Broderick. Geotech. Special Pub. 76, ASCE, 19.

REFERENCES

1. Vidal, H. 1966. La terre Armée. *Anales de l'institut Technique du Bâtiment et des Travaus Publiques*, France, July–August, 888.
2. Binquet, J., and K. L. Lee. 1975. Bearing capacity tests on reinforced earth mass. *J. Geotech. Eng. Div.*, ASCE, 101(12): 1241.
3. Binquet, J., and K. L. Lee. 1975. Bearing capacity analysis of reinforced earth slabs. *J. Geotech. Eng. Div.*, ASCE, 101(12): 1257.
4. Omar, M. T., B. M. Das, S. C. Yen, V. K. Puri, and E. E. Cook. 1993. Ultimate bearing capacity of rectangular foundations on geogrid-reinforced sand. *Geotech. Testing J.*, ASTM, 16(2): 246.
5. Guido, V. A., J. D. Knueppel, and M. A. Sweeney. 1987. Plate load tests on geogrid-reinforced earth slabs, in *Proc., Geosynthetics 1987*, 216.
6. Akinmusuru, J. O., and J. A. Akinbolade. 1981. Stability of loaded footings on reinforced soil. *J. Geotech. Engg. Div.*, ASCE, 107:819.
7. Yetimoglu, T., J. T. H. Wu, and A. Saglamer. 1994. Bearing capacity of rectangular footings on geogrid-reinforced sand. *J. Geotech. Eng.*, ASCE, 120(12): 2083.

8. Schlosser, F., H. M. Jacobsen, and I. Juran. 1983. Soil reinforcement—general report. *Proc. VIII European Conf. Soil Mech. Found. Engg.* Helsinki, Balkema, 83.

9. Adams, M. T., and J. G. Collin. 1997. Large model spread footing load tests on geosynthetic reinforced soil foundation. *J. Geotech. Geoenviron. Eng.,* ASCE, 123(1): 66.

10. Shin, E. C., and B. M. Das. 2000. Experimental study of bearing capacity of a strip foundation on geogrid-reinforced sand. *Geosynthetics Intl.* 7(1): 59.

11. Huang, C. C., and L. K. Hong. 2000. Ultimate bearing capacity and settlement of footings on reinforced sandy ground. *Soils and Foundations* 49(5): 65.

12. Huang, C. C., and F. Tatsuoka. 1990. Bearing capacity of reinforced horizontal sandy ground. *Geotextiles and Geomembranes* 9: 51.

13. Takemura, J., M. Okamura, N. Suesmasa, and T. Kimura. 1992. Bearing capacity and deformations of sand reinforced with geogrids. *Proc., Int. Symp. Earth Reinforcement Practice,* Fukuoka, Japan, 695.

14. Khing, K. H., B. M. Das, V. K. Puri, E. E. Cook, and S. C. Yen. 1992. Bearing capacity of two closely spaced strip foundations on geogrid-reinforced sand. *Proc., Int. Symp. Earth Reinforcement Practice,* Fukuoka, Japan, 619.

15. Huang, C. C., and F. Y. Meng. 1997. Deep footing and wide-slab effects on reinforced sandy ground. *J. Geotech. Geoenviron. Eng.,* ASCE, 123(1): 30.

16. Patra, C. R., B. M. Das, M. Bohi, and E. C. Shin. 2006. Eccentrically loaded strip foundation on geogrid-reinforced sand. *Geotextiles and Geomembranes* 24: 254.

17. Das, B. M. 1998. Dynamic loading on foundation on reinforced soil, in *Geosynthetics in foundation reinforcement and erosion control systems,* eds. J. J. Bowders, H. B. Scranton, and G. P. Broderick. Geotech. Special Pub. 76, ASCE, 19.

8 Uplift Capacity of Shallow Foundations

8.1 INTRODUCTION

Foundations and other structures may be subjected to uplift forces under special circumstances. For those foundations, during the design process it is desirable to apply a sufficient factor of safety against failure by uplift. During the last 40 or so years, several theories have been developed to estimate the ultimate uplift capacity of foundations embedded in sand and clay soils, and some of those theories are detailed in this chapter. The chapter is divided into two major parts: foundations in granular soil and foundations in saturated clay soil ($\phi = 0$).

Figure 8.1 shows a shallow foundation of width B and depth of embedment D_f. The ultimate uplift capacity of the foundation Q_u can be expressed as

$$Q_u = \text{frictional resistance of soil along the failure surface}$$

$$+ \text{ weight of soil in the failure zone and the foundation} \qquad (8.1)$$

If the foundation is subjected to an uplift load of Q_u, the failure surface in the soil for relatively small D_f/B values will be of the type shown in Figure 8.1. The intersection of the failure surface at the ground level will make an angle α with the horizontal. However, the magnitude of α will vary with the relative density of compaction in the case of sand, and with the consistency in the case of clay soils.

When the failure surface in soil extends up to the ground surface at ultimate load, it is defined as a *shallow foundation under uplift*. For larger values of D_f/B, failure takes place around the foundation and the failure surface does not extend to the ground surface. These are called *deep foundations under uplift*. The *embedment ratio* D_f/B at which a foundation changes from shallow to deep condition is referred to as the critical embedment ratio $(D_f/B)_{cr}$. In sand the magnitude of $(D_f/B)_{cr}$ can vary from 3 to about 11, and in saturated clay it can vary from 3 to about 7.

8.2 FOUNDATIONS IN SAND

During the last 40 years, several theoretical and semi-empirical methods have been developed to predict the net ultimate uplifting load of continuous, circular, and rectangular foundations embedded in sand. Some of these theories are briefly described in the following sections.

8.2.1 BALLA'S THEORY

Based on the results of several model and field tests conducted in dense soil, Balla[1] established that, for *shallow circular foundations*, the failure surface in soil will be as

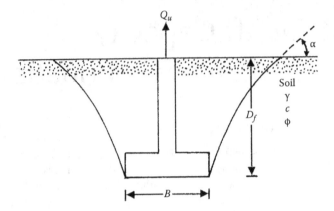

FIGURE 8.1 Shallow foundation subjected to uplift.

shown in Figure 8.2. Note from the figure that aa' and bb' are arcs of a circle. The angle α is equal to $45 - \phi/2$. The radius of the circle, of which aa' and bb' are arcs, is equal to

$$r = \frac{D_f}{\sin\left(45 + \frac{\phi}{2}\right)} \tag{8.2}$$

As mentioned before, the ultimate uplift capacity of the foundation is the sum of two components: (a) the weight of the soil and the foundation in the failure zone and (b) the shearing resistance developed along the failure surface. Thus, assuming that the unit weight of soil and the foundation material are approximately the same,

$$Q_u = D_f^3 \gamma \left[F_1\left(\phi, \frac{D_f}{B} \right) + F_3\left(\phi, \frac{D_f}{B} \right) \right] \tag{8.3}$$

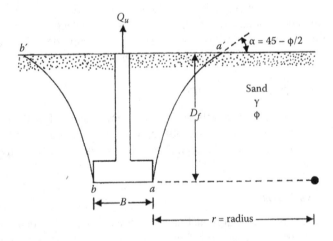

FIGURE 8.2 Balla's theory for shallow circular foundations.

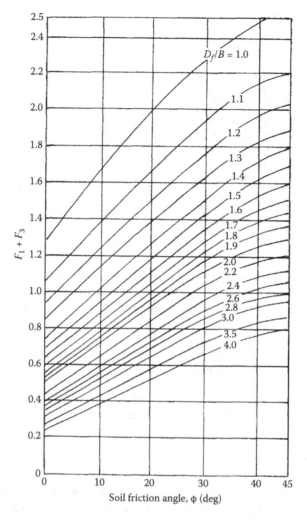

FIGURE 8.3 Variation of $F_1 + F_3$ [equation (8.3)].

where

γ = unit weight of soil
ϕ = soil friction angle
B = diameter of the circular foundation

The sums of the functions $F_1(\phi, D_f/B)$ and $F_3(\phi, D_f/B)$ developed by Balla[1] are plotted in Figure 8.3 for various values of the soil friction angle ϕ and the embedment ratio, D_f/B.

In general, Balla's theory is in good agreement with the uplift capacity of shallow foundations embedded in dense sand at an embedment ratio of $D_f/B \leq 5$. However, for foundations located in loose and medium sand, the theory overestimates the

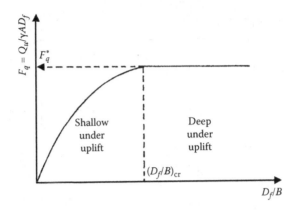

FIGURE 8.4 Nature of variation of F_q with D_f/B.

ultimate uplift capacity. The main reason Balla's theory overestimates the ultimate uplift capacity for $D_f/B >$ about 5 even in dense sand is because it is essentially a deep foundation condition, and the failure surface does not extend to the ground surface.

The simplest procedure to determine the embedment ratio at which the deep foundation condition is reached may be determined by plotting the nondimensional breakout factor F_q against D_f/B as shown in Figure 8.4. The breakout factor is derived as

$$F_q = \frac{Q_u}{\gamma A D_f} \tag{8.4}$$

where
    A = area of the foundation.

The breakout factor increases with D_f/B up to a maximum value of $F_q = F_q^*$ at $D_f/B = (D_f/B)_{cr}$. For $D_f/B > (D_f/B)_{cr}$ the breakout factor remains practically constant (that is, F_q^*).

8.2.2 THEORY OF MEYERHOF AND ADAMS

One of the most rational methods for estimating the ultimate uplift capacity of a shallow foundation was proposed by Meyerhof and Adams,[2] and it is described in detail in this section. Figure 8.5 shows a continuous foundation of width B subjected to an uplifting force. The ultimate uplift capacity per unit length of the foundation is equal to Q_u. At ultimate load the failure surface in soil makes an angle α with the horizontal. The magnitude of α depends on several factors, such as the relative density of compaction and the angle of friction of the soil, and it varies between $90° - 1/3\ \phi$ and $90° - 2/3\ \phi$. Let us consider the free body diagram of the zone $abcd$. For stability consideration, the following forces *per unit length of the foundation* need to be considered: (a) the weight of the soil and concrete W and (b) the passive force P_p' per unit length along the faces ad and bc. The force P_p' is inclined at an

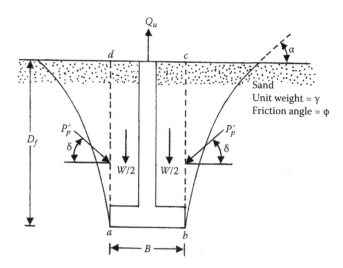

FIGURE 8.5 Continuous foundation subjected to uplift.

angle δ to the horizontal. For an average value of $\alpha = 90 - \phi/2$, the magnitude of δ is about 2/3 ϕ.

If we assume that the unit weights of soil and concrete are approximately the same, then

$$W = \gamma D_f B$$

$$P'_p = \frac{P'_h}{\cos\delta} = \left(\frac{1}{2}\right)\left(\frac{1}{\cos\delta}\right)(K_{ph}\gamma D_f^2) \tag{8.5}$$

where

P'_h = horizontal component of the passive force
K_{ph} = horizontal component of the passive earth pressure coefficient

Now, for equilibrium, summing the vertical components of all forces,

$$\sum F_v = 0$$

$$Q_u = W + 2P'_p \sin\delta$$

$$Q_u = W + 2(P'_p \cos\delta)\tan\delta$$

$$Q_u = W + 2P'_h \tan\delta$$

or

$$Q_u = W + 2\left(\tfrac{1}{2}K_{ph}\gamma D_f^2\right)\tan\delta = W + K_{ph}\gamma D_f^2 \tan\delta \tag{8.6}$$

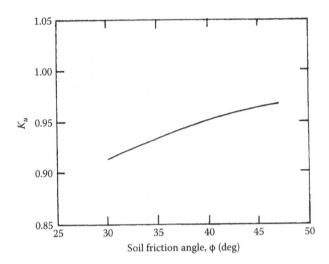

FIGURE 8.6 Variation of K_u.

The passive earth pressure coefficient based on the curved failure surface for $\delta = 2/3\ \phi$ can be obtained from Caquot and Kerisel.[3] Furthermore, it is convenient to express $K_{ph} \tan \delta$ in the form

$$K_u \tan \phi = K_{ph} \tan \delta \qquad (8.7)$$

Combining equations (8.6) and (8.7),

$$Q_u = W + K_u \gamma D_f^2 \tan \phi \qquad (8.8)$$

where
$\quad K_u$ = nominal uplift coefficient

The variation of the nominal uplift coefficient K_u with the soil friction angle ϕ is shown in Figure 8.6. It falls within a narrow range and may be taken as equal to 0.95 for all values of ϕ varying from 30° to about 48°. The ultimate uplift capacity can now be expressed in a nondimensional form (that is, the breakout factor, F_q) as defined in equation (8.4).[4] Thus, for a continuous foundation, the breakout factor per unit length is

$$F_q = \frac{Q_u}{\gamma A D_f}$$

or

$$F_q = \frac{W + K_u \gamma D_f^2 \tan \phi}{W} = 1 + K_u \left(\frac{D_f}{B} \right) \tan \phi \qquad (8.9)$$

For circular foundations, equation (8.8) can be modified to the form

$$Q_u = W + \frac{\pi}{2} S_F \gamma B D_f^2 K_u \tan \phi \qquad (8.10)$$

$$W \approx \frac{\pi}{4} B^2 D_f \gamma \qquad (8.11)$$

where
S_F = shape factor
B = diameter of the foundation

The shape factor can be expressed as

$$S_F = 1 + m \left(\frac{D_f}{B} \right) \qquad (8.12)$$

where
m = coefficient that is a function of the soil friction angle ϕ

Thus, combining equations (8.10), (8.11), and (8.12) we obtain

$$Q_u = \frac{\pi}{4} B^2 D_f \gamma + \frac{\pi}{2} \left[1 + m \left(\frac{D_f}{B} \right) \right] \gamma B D_f^2 K_u \tan \phi \qquad (8.13)$$

The breakout factor F_q can be given as

$$F_q = \frac{Q_u}{\gamma A D_f} = \frac{\frac{\pi}{4} B^2 D_f \gamma + \frac{\pi}{2} \left[1 + m \left(\frac{D_f}{B} \right) \right] \gamma B D_f^2 K_u \tan \phi}{\gamma \left(\frac{\pi}{4} B^2 \right) D_f}$$

$$= 1 + 2 \left[1 + m \left(\frac{D_f}{B} \right) \right] \left(\frac{D_f}{B} \right) K_u \tan \phi \qquad (8.14)$$

For rectangular foundations having dimensions of $B \times L$, the ultimate capacity can also be expressed as

$$Q_u = W + \gamma D_f^2 (2 S_F B + L - B) K_u \tan \phi \qquad (8.15)$$

The preceding equation was derived with the assumption that the two end portions of length $B/2$ are governed by the shape factor S_F, while the passive pressure along the central portion of length $L - B$ is the same as the continuous foundation. In equation (8.15),

$$W \approx \gamma B L D_f \qquad (8.16)$$

and

$$S_F = 1 + m \left(\frac{D_f}{B} \right) \qquad (8.17)$$

TABLE 8.1
Variation of m [Equation (8.12)]

Soil Friction Angle ϕ	m
20	0.05
25	0.1
30	0.15
35	0.25
40	0.35
45	0.5
48	0.6

Thus,

$$Q_u = \gamma BLD_f + \gamma D_f^2 \left\{ 2\left[1+m\left(\frac{D_f}{B}\right)\right]B+L-B\right\} K_u \tan\phi \qquad (8.18)$$

The breakout factor F_q can now be determined as

$$F_q = \frac{Q_u}{\gamma BLD_f} \qquad (8.19)$$

Combining equations (8.18) and (8.19), we obtain[4]

$$F_q = 1+\left\{\left[1+2m\left(\frac{D_f}{B}\right)\right]\left(\frac{B}{L}\right)+1\right\}\left(\frac{D_f}{B}\right)K_u \tan\phi \qquad (8.20)$$

The coefficient m given in equation (8.12) was determined from experimental observations[2] and its values are given in Table 8.1. As shown in Figure 8.4, the breakout factor F_q increases with D_f/B to a maximum value of F_q^* at $(D_f/B)_{cr}$ and remains constant thereafter. Based on experimental observations, Meyerhof and Adams[2] recommended the variation of $(D_f/B)_{cr}$ for square and circular foundations with soil friction angle ϕ and this is shown in Figure 8.7.

Thus, for a given value of ϕ for square ($B = L$) and circular (diameter = B) foundations, we can substitute m (Table 8.1) into equations (8.14) and (8.20) and calculate the breakout factor F_q variation with embedment ratio D_f/B. The maximum value of $F_q = F_q^*$ will be attained at $D_f/B = (D_f/B)_{cr}$. For $D_f/B > (D_f/B)_{cr}$, the breakout factor will remain constant as F_q^*. The variation of F_q with D_f/B for various values of ϕ made in this manner is shown in Figure 8.8. Figure 8.9 shows the variation of the maximum breakout factor F_q^* for deep square and circular foundations with the soil friction angle ϕ.

Laboratory experimental observations have shown that the critical embedment ratio (for a given soil friction angle ϕ) increases with the L/B ratio. For a given value of ϕ, Meyerhof[5] indicated that

$$\frac{\left(\frac{D_f}{B}\right)_{cr\text{-continuous}}}{\left(\frac{D_f}{B}\right)_{cr\text{-square}}} \approx 1.5 \qquad (8.21)$$

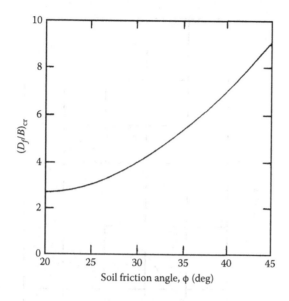

FIGURE 8.7 Variation of $(D_f/B)_{cr}$ for square and circular foundations.

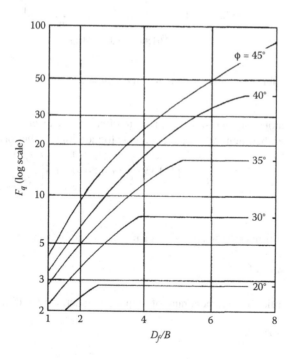

FIGURE 8.8 Plot of F_q for square and circular foundations [equations (8.14) and (8.20)].

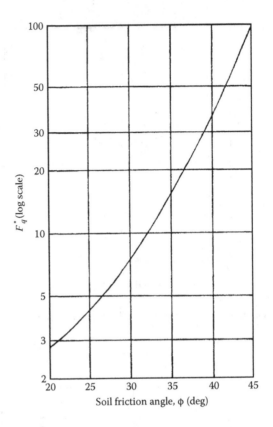

FIGURE 8.9 F_q^* for deep square and circular foundations.

Based on laboratory model test results, Das and Jones[6] gave an empirical relationship for the critical embedment ratio of rectangular foundations in the form

$$\left(\frac{D_f}{B}\right)_{\text{cr-R}} = \left(\frac{D_f}{B}\right)_{\text{cr-S}}\left[0.133\left(\frac{L}{B}\right)+0.867\right] \leq 1.4\left(\frac{D_f}{B}\right)_{\text{cr-S}} \qquad (8.22)$$

where

$\left(\dfrac{D_f}{B}\right)_{\text{cr-R}}$ = critical embedment ratio of a rectangular foundation with dimensions

      of $L \times B$

$\left(\dfrac{D_f}{B}\right)_{\text{cr-S}}$ = critical embedment ratio of a square foundation with dimensions

      of $B \times B$

Using equation (8.22) and the $(D_f/B)_{\text{cr-S}}$ values given in Figure 8.7, the magnitude of $(D_f/B)_{\text{cr-R}}$ for a rectangular foundation can be estimated. These values of $(D_f/B)_{\text{cr-R}}$

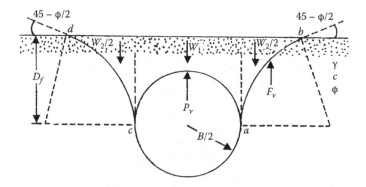

FIGURE 8.10 Vesic's theory of expansion of cavities.

can be substituted into equation (8.20) to determine the variation of $F_q = F^*$ with the soil friction angle ϕ.

8.2.3 THEORY OF VESIC

Vesic[7] studied the problem of an explosive point charge expanding a spherical cavity close to the surface of a semi-infinite, homogeneous, isotropic solid (in this case, the soil). Referring to Figure 8.10, it can be seen that if the distance D_f is small enough there will be an ultimate pressure p_o that will shear away the soil located above the cavity. At that time, the diameter of the spherical cavity is equal to B. The slip surfaces ab and cd will be tangent to the spherical cavity at a and c. At points b and d they make an angle $\alpha = 45 - \phi/2$. For equilibrium, summing the components of forces in the vertical direction we can determine the ultimate pressure p_o in the cavity. Forces that will be involved are

1. Vertical component of the force inside the cavity P_V
2. Effective self-weight of the soil $W = W_1 + W_2$
3. Vertical component of the resultant of internal forces F_V

For a $c - \phi$ soil, we can thus determine that

$$p_o = c\bar{F}_c + \gamma D_f \bar{F}_q \qquad (8.23)$$

where

$$\bar{F} = 1.0 - \frac{2}{3}\left[\frac{\left(\dfrac{B}{2}\right)}{D_f}\right] + A_1\left[\frac{D_f}{\left(\dfrac{B}{2}\right)}\right] + A_2\left[\frac{D_f}{\left(\dfrac{B}{2}\right)}\right]^2 \qquad (8.24)$$

$$\bar{F}_c = A_2\left[\frac{D_f}{\left(\dfrac{B}{2}\right)}\right] + A_4\left[\frac{D_f}{\left(\dfrac{B}{2}\right)}\right] \qquad (8.25)$$

where
A_1, A_2, A_3, A_4 = functions of the soil friction angle ϕ

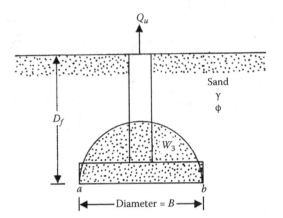

FIGURE 8.11 Cavity expansion theory applied to circular foundation uplift.

For granular soils $c = 0$, so

$$p_o = \gamma D_f \bar{F}_q \tag{8.26}$$

Vesic[8] applied the preceding concept to determine the ultimate uplift capacity of shallow circular foundations. In Figure 8.11 consider that the circular foundation *ab* with a diameter *B* is located at a depth D_f below the ground surface. Assuming that the unit weight of the soil and the unit weight of the foundation are approximately the same, if the hemispherical cavity above the foundation (that is, *ab*) is filled with soil, it will have a weight of

$$W_3 = \frac{2}{3}\pi\left(\frac{B}{2}\right)^3\gamma \tag{8.27}$$

This weight of soil will increase the pressure by p_1, or

$$p_1 = \frac{W_3}{\pi\left(\frac{B}{2}\right)^2} = \frac{\left(\frac{2}{3}\right)\pi\left(\frac{B}{2}\right)^3\gamma}{\pi\left(\frac{B}{2}\right)^2} = \frac{2}{3}\pi\left(\frac{B}{2}\right)$$

If the foundation is embedded in a cohesionless soil ($c = 0$), the pressure p_1 should be added to equation (8.26) to obtain the force per unit area of the anchor q_u needed for complete pullout. Thus,

$$q_u = \frac{Q_u}{A} = \frac{Q_u}{\frac{\pi}{2}(B)^2} = p_o + p_1 = \gamma D_f \bar{F}_q + \frac{2}{3}\gamma\left(\frac{B}{2}\right) = \gamma D_f\left[\bar{F}_q + \frac{\left(\frac{2}{3}\right)\left(\frac{B}{2}\right)}{D_f}\right] \tag{8.28}$$

or

$$q_u = \frac{Q_u}{A} = \gamma D_f\left\{1 + A_1\left[\frac{D_f}{\left(\frac{B}{2}\right)}\right] + A_2\left[\frac{D_f}{\left(\frac{B}{2}\right)}\right]^2\right\} = \gamma D_f \underbrace{F_q}_{\substack{\text{breakout}\\\text{factor}}} \tag{8.29}$$

TABLE 8.2
Vesic's Breakout Factor F_q for Circular Foundations

Soil Friction Angle ϕ (deg)	D_f/B				
	0.5	1.0	1.5	2.5	5.0
0	1.0	1.0	1.0	1.0	1.0
10	1.18	1.37	1.59	2.08	3.67
20	1.36	1.75	2.20	3.25	6.71
30	1.52	2.11	2.79	4.41	9.89
40	1.65	2.41	3.30	5.43	13.0
50	1.73	2.61	3.56	6.27	15.7

The variations of the breakout factor F_q for *shallow circular foundations* are given in Table 8.2 and Figure 8.12. In a similar manner, Vesic determined the variation of the breakout factor F_q for *shallow continuous foundations* using the analogy of expansion of long cylindrical cavities. These values are given in Table 8.3 and are also plotted in Figure 8.13.

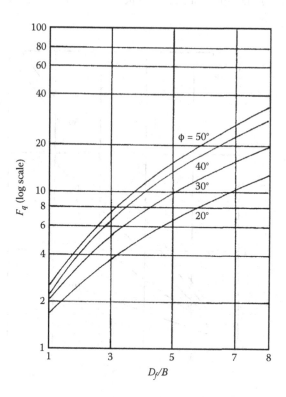

FIGURE 8.12 Vesic's breakout factor F_q for shallow circular foundations.

TABLE 8.3
Vesic's Breakout Factor F_q for Continuous Foundations

Soil Friction Angle ϕ (deg)	D_f/B				
	0.5	1.0	1.5	2.5	5.0
0	1.0	1.0	1.0	1.0	1.0
10	1.09	1.16	1.25	1.42	1.83
20	1.17	1.33	1.49	1.83	2.65
30	1.24	1.47	1.71	2.19	3.38
40	1.30	1.58	1.87	2.46	3.91
50	1.32	1.64	2.04	2.60	4.20

8.2.4 Saeddy's Theory

A theory for the ultimate uplift capacity of *circular foundations* embedded in sand was proposed by Saeedy[9] in which the trace of the failure surface was assumed to be an arc of a logarithmic spiral. According to this solution, for shallow foundations the failure surface extends to the ground surface. However, for deep foundations (that is, $D_f > D_{f(cr)}$) the failure surface extends only to a distance of $D_{f(cr)}$ above the

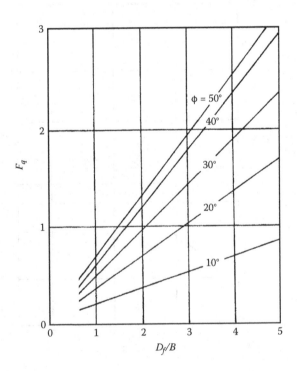

FIGURE 8.13 Vesic's breakout factor F_q for shallow continuous foundations.

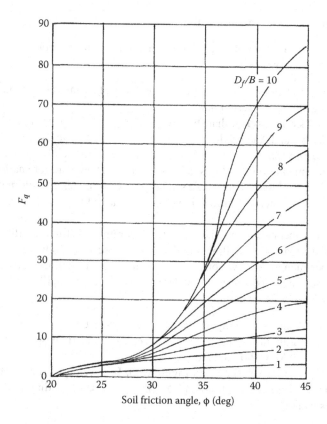

FIGURE 8.14 Plot of F_q based on Saeedy's theory.

foundation. Based on this analysis, Saeedy[9] proposed the ultimate uplift capacity in a nondimensional form $(Q_u/\gamma B^2 D_f)$ for various values of ϕ and the D_f/B ratio. The author converted the solution into a plot of breakout factor $F_q = Q_u/\gamma A D_f$ (A = area of the foundation) versus the soil friction angle ϕ as shown in Figure 8.14. According to Saeedy, during the foundation uplift the soil located above the anchor gradually becomes compacted, in turn increasing the shear strength of the soil and hence the ultimate uplift capacity. For that reason, he introduced an empirical *compaction factor* μ, which is given in the form

$$\mu = 1.044 D_r + 0.44 \tag{8.30}$$

where
  D_r = relative density of sand

Thus, the actual ultimate capacity can be expressed as

$$Q_{u(\text{actual})} = (F_q \gamma A D_f)\mu \tag{8.31}$$

8.2.5 Discussion of Various Theories

Based on the various theories presented in the preceding sections, we can make some general observations:

1. The only theory that addresses the problem of rectangular foundations is that given by Meyerhof and Adams.[2]
2. Most theories assume that shallow foundation conditions exist for $D_f/B \leq 5$. Meyerhof and Adams' theory provides a critical embedment ratio $(D_f/B)_{cr}$ for *square and circular* foundations as a function of the soil friction angle.
3. Experimental observations generally tend to show that, for shallow foundations in loose sand, Balla's theory[1] overestimates the ultimate uplift capacity. Better agreement, however, is obtained for foundations in dense soil.
4. Vesic's theory[8] is, in general, fairly accurate for estimating the ultimate uplift capacity of shallow foundations in loose sand. However, laboratory experimental observations have shown that, for shallow foundations in dense sand, this theory can underestimate the actual uplift capacity by as much as 100% or more.

Figure 8.15 shows a comparison of some published laboratory experimental results for the ultimate uplift capacity of *circular* foundations with the theories of Balla, Vesic, and Meyerhof and Adams. Table 8.4 gives the references to the laboratory experimental curves shown in Figure 8.15. In developing the theoretical plots for $\phi = 30°$ (loose sand condition) and $\phi = 45°$ (dense sand condition), the following procedures were used:

1. According to Balla's theory,[1] from equation (8.3) for circular foundations,

$$Q_u = D_f^3 \gamma (F_1 + F_3)$$

So,

$$F_1 + F_3 = \frac{Q_u}{\gamma D_f^3} = \frac{\left(\frac{\pi}{4} B^2\right) Q_u}{\gamma D_f^3 \left(\frac{\pi}{4} B^2\right)} = \frac{\left[\left(\frac{\pi}{4}\right)\left(\frac{B}{D_f}\right)^2\right] Q_u}{\gamma D_f A}$$

or

$$F_q = \frac{Q_u}{\gamma A D_f} = \frac{F_1 + F_3}{\left(\frac{\pi}{4}\right)\left(\frac{B}{D_f}\right)^2} \tag{8.32}$$

So, for a given soil friction angle, the sum of $F_1 + F_3$ was obtained from Figure 8.3, and the breakout factor was calculated for various values of D_f/B. These values are plotted in Figure 8.15.

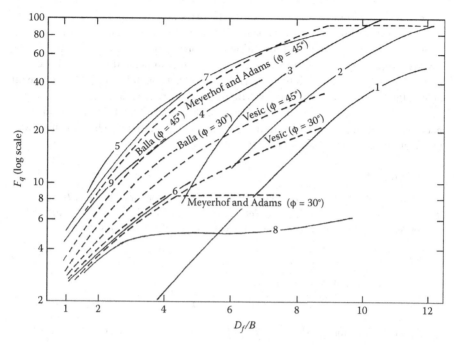

FIGURE 8.15 Comparison of theories with laboratory experimental results for circular foundations.

2. For Vesic's theory,[8] the variations of F_q versus D_f/B for circular foundations are given in Table 8.2. These values of F_q are also plotted in Figure 8.15.

3. The breakout factor relationship for circular foundations based on Meyerhof and Adams' theory[3] is given in equation (8.14). Using $K_u \approx 0.95$, the variations of F_q with D_f/B were calculated, and they are also plotted in Figure 8.15.

TABLE 8.4
References to Laboratory Experimental Curves Shown in Figure 8.15

Curve	Reference	Circular Foundation Diameter B (mm)	Soil Properties
1	Baker and Kondner[10]	25.4	$\phi = 42°$; $\gamma = 17.61$ kN/m³
2	Baker and Kondner[10]	38.1	$\phi = 42°$; $\gamma = 17.61$ kN/m³
3	Baker and Kondner[10]	50.8	$\phi = 42°$; $\gamma = 17.61$ kN/m³
4	Baker and Kondner[10]	76.2	$\phi = 42°$; $\gamma = 17.61$ kN/m³
5	Sutherland[11]	38.1–152.4	$\phi = 45°$
6	Sutherland[11]	38.1–152.4	$\phi = 31°$
7	Esquivel-Diaz[12]	76.2	$\phi \approx 43°$; $\gamma = 14.81–15.14$ kN/m³
8	Esquivel-Diaz[12]	76.2	$\phi = 33°$; $\gamma = 12.73–12.89$ kN/m³
9	Balla[1]	61–119.4	Dense sand

Based on the comparison between the theories and the laboratory experimental results shown in Figure 8.15, it appears that Meyerhof and Adams' theory[2] is more applicable to a wide range of foundations and provides as good an estimate as any for the ultimate uplift capacity. So this theory is recommended for use. However, it needs to be kept in mind that the majority of the experimental results presently available in the literature for comparison with the theory are from laboratory model tests. When applying these results to the design of an actual foundation, the *scale effect* needs to be taken into consideration. For that reason, a judicious choice is necessary in selecting the value of the soil friction angle ϕ.

EXAMPLE 8.1

Consider a circular foundation in sand. Given, for the foundation: diameter $B = 1.5$ m; depth of embedment $D_f = 1.5$ m. Given, for the sand: unit weight $\gamma = 17.4$ kN/m³; friction angle $\phi = 35°$. Using Balla's theory, calculate the ultimate uplift capacity.

Solution

From equation (8.3),

$$Q_u = D_f^3 \gamma (F_1 + F_3)$$

From Figure 8.3 for $\phi = 35°$ and $D_f/B = 1.5/1.5 = 1$, the magnitude of $F_1 + F_3 \approx 2.4$. So,

$$Q_u = (1.5)^3 (17.4)(2.4) = \textbf{140.9 kN}$$

EXAMPLE 8.2

Redo Example 8.1 problem using Vesic's theory.

Solution

From equation (8.29),

$$Q_u = A\gamma D_f F_q$$

From Figure 8.12 for $\phi = 35°$ and $D_f/B = 1$, F_q is about 2.2. So,

$$Q_u = \left[\left(\frac{\pi}{4} \right)(1.5)^2 \right](17.4)(1.5)(2.2) = \textbf{101.5 kN}$$

EXAMPLE 8.3

Redo Example 8.1 problem using Meyerhof and Adams' theory.

Solution

From equation (8.14),

$$F_q = 1 + 2\left[1 + m\left(\frac{D_f}{B}\right)\right]\left(\frac{D_f}{B}\right)K_u \tan\phi$$

For $\phi = 35°$, $m = 0.25$ (Table 8.1). So,

$$F_q = 1 + 2[1 + (0.25)(1)](1)(0.95)(\tan 35) = 2.66$$

So,

$$Q_u = F_q \gamma A D_f = (2.66)(17.4)\left[\left(\frac{\pi}{4}\right)(1.5)^2\right](1.5) = \mathbf{122.7\ kN}$$

8.3 FOUNDATIONS IN SATURATED CLAY (ϕ = 0 CONDITION)

8.3.1 ULTIMATE UPLIFT CAPACITY—GENERAL

Theoretical and experimental research results presently available for determining the ultimate uplift capacity of foundations embedded in saturated clay soil are rather limited. In the following sections, the results of some of the existing studies are reviewed.

Figure 8.16 shows a shallow foundation in saturated clay. The depth of the foundation is D_f, and the width of the foundation is B. The undrained shear strength and the unit weight of the soil are c_u and γ, respectively. If we assume that the unit weights of the foundation material and the clay are approximately the same, then the ultimate uplift capacity can be expressed as[8]

$$Q_u = A(\gamma D_f + c_u F_c) \tag{8.33}$$

where
    A = area of the foundation
    F_c = breakout factor
    γ = saturated unit weight of the soil

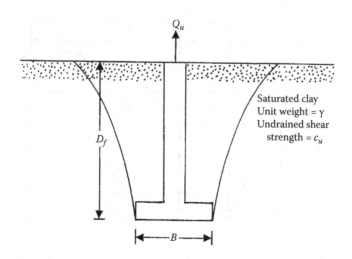

FIGURE 8.16 Shallow foundation in saturated clay subjected to uplift.

TABLE 8.5
Variation of F_c ($\phi = 0$ Condition)

Foundation Type	D_f/B				
	0.5	1.0	1.5	2.5	5.0
Circular (diameter = B)	1.76	3.80	6.12	11.6	30.3
Continuous (width = B)	0.81	1.61	2.42	4.04	8.07

8.3.2 Vesic's Theory

Using the analogy of the expansion of cavities, Vesic[8] presented the theoretical variation of the breakout factor F_c (for $\phi = 0$ condition) with the embedment ratio D_f/B, and these values are given in Table 8.5. A plot of these same values of F_c against D_f/B is also shown in Figure 8.17. Based on the laboratory model test results available at the present time, it appears that Vesic's theory gives a closer estimate only for shallow foundations embedded in softer clay.

In general, the breakout factor increases with the embedment ratio up to a maximum value and remains constant thereafter, as shown in Figure 8.18. The maximum value of $F_c = F_c^*$ is reached at $D_f/B = (D_f/B)_{cr}$. Foundations located at $D_f/B > (D_f/B)_{cr}$

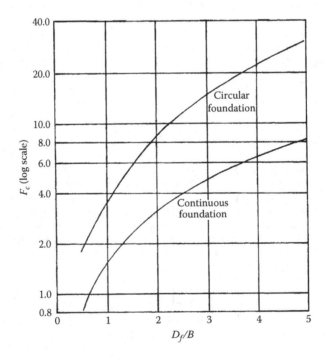

FIGURE 8.17 Vesic's breakout factor F_c.

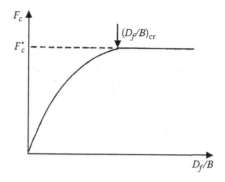

FIGURE 8.18 Nature of variation of F_c with D_f/B.

are referred to as deep foundations for uplift capacity consideration. For these foundations at ultimate uplift load, local shear failure in soil located around the foundation takes place. Foundations located at $D_f/B \leq (D_f/B)_{cr}$ are shallow foundations for uplift capacity consideration.

8.3.3 MEYERHOF'S THEORY

Based on several experimental results, Meyerhof[5] proposed the following relationship:

$$Q_u = A(\gamma D_f + F_c c_u)$$

For circular and square foundations,

$$F_c = 1.2\left(\frac{D_f}{B}\right) \leq 9 \tag{8.34}$$

and for *strip* foundations,

$$F_c = 0.6\left(\frac{D_f}{B}\right) \leq 8 \tag{8.35}$$

The preceding two equations imply that the critical embedment ratio $(D_f/B)_{cr}$ is about 7.5 for square and circular foundations and about 13.5 for strip foundations.

8.3.4 MODIFICATIONS TO MEYERHOF'S THEORY

Das[13] compiled a number of laboratory model test results on circular foundations in saturated clay with c_u varying from 5.18 kN/m² to about 172.5 kN/m². Figure 8.19 shows the average plots of F_c versus D_f/B obtained from these studies along with the critical embedment ratios. From Figure 8.19 it can be seen that, for shallow foundations,

$$F_c \approx n\left(\frac{D_f}{B}\right) \leq 8 \text{ to } 9 \tag{8.36}$$

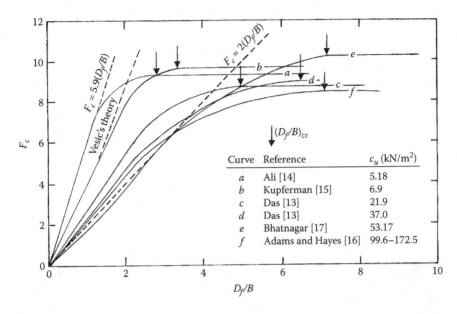

FIGURE 8.19 Variation of F_c with D_f/B from various experimental observations—circular foundation; diameter = B.

where

n = a constant

The magnitude of n varies from 5.9 to 2.0 and is a function of the undrained cohesion. Since n is a function of c_u and $F_c = F_c^*$ is about eight to nine in all cases, it is obvious that the critical embedment ratio $(D_f/B)_{cr}$ will be a function of c_u.

Das[13] also reported some model test results with square and rectangular foundations. Based on these tests, it was proposed that

$$\left(\frac{D_f}{B}\right)_{cr\text{-}S} = 0.107c_u + 2.5 \le 7 \tag{8.37}$$

where

$\left(\dfrac{D_f}{B}\right)_{cr\text{-}S}$ = critical embedment ratio of square foundations (or circular foundations)

c_u = undrained cohesion, in kN/m^2

It was also observed by Das[18] that

$$\left(\frac{D_f}{B}\right)_{cr\text{-}R} = \left(\frac{D_f}{B}\right)_{cr\text{-}S}\left[0.73 + 0.27\left(\frac{L}{B}\right)\right] \le 1.55\left(\frac{D_f}{B}\right)_{cr\text{-}S} \tag{8.38}$$

where

$$\left(\frac{D_f}{B}\right)_{\text{cr-R}} = \text{critical embedment ratio of rectangular foundations}$$

$$L = \text{length of foundation}$$

Based on the above findings, Das[18] proposed an empirical procedure to obtain the breakout factors for shallow and deep foundations. According to this procedure, α' and β' are two nondimensional factors defined as

$$\alpha' = \frac{\dfrac{D_f}{B}}{\left(\dfrac{D_f}{B}\right)_{\text{cr}}} \tag{8.39}$$

and

$$\beta' = \frac{F_c}{F_c^*} \tag{8.40}$$

For a given foundation, the critical embedment ratio can be calculated using equations (8.37) and (8.38). The magnitude of F_c^* can be given by the following empirical relationship:

$$F_{c\text{-R}}^* = 7.56 + 1.44\left(\frac{B}{L}\right) \tag{8.41}$$

where

$$F_{c-R}^* = \text{breakout factor for deep rectangular foundations}$$

Figure 8.20 shows the experimentally derived plots (upper limit, lower limit, and average of β' and α'). Following is a step-by-step procedure to estimate the ultimate uplift capacity.

1. Determine the representative value of the undrained cohesion c_u.
2. Determine the critical embedment ratio using equations (8.37) and (8.38).
3. Determine the D_f/B ratio for the foundation.
4. If $D_f/B > (D_f/B)_{\text{cr}}$ as determined in step 2, it is a deep foundation. However, if $D_f/B \le (D_f/B)_{\text{cr}}$, it is a shallow foundation.
5. For $D_f/B > (D_f/B)_{\text{cr}}$,

$$F_c = F_c^* = 7.56 + 1.44\left(\frac{B}{L}\right)$$

Thus,

$$Q_u = A\left\{\left[7.56 + 1.44\left(\frac{B}{L}\right)\right]c_u + \gamma D_f\right\} \tag{8.42}$$

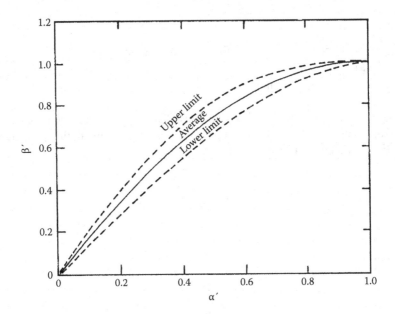

FIGURE 8.20 Plot of β' versus α'.

where

    A = area of the foundation

    6. For $D_f/B \le (D_f/B)_{cr}$,

$$Q_u = A(\beta' F_c^* c_u + \gamma D_f) = A\left\{\beta'\left[7.56 + 1.44\left(\frac{B}{L}\right)\right]c_u + \gamma D_f\right\} \tag{8.43}$$

The value of β' can be obtained from the average curve of Figure 8.20. The procedure outlined above gives fairly good results in estimating the net ultimate capacity of foundations.

EXAMPLE 8.4

A rectangular foundation in saturated clay measures 1.5 m × 3 m. Given: D_f = 1.8 m; c_u = 52 kN/m²; γ = 18.9 kN/m³. Estimate the ultimate uplift capacity.

Solution

From equation (8.37),

$$\left(\frac{D_f}{B}\right)_{cr-S} = 0.107 c_u + 2.5 = (0.107)(52) + 2.5 = 8.06$$

So use $(D_f/B)_{cr-S}$ = 7. Again, from equation (8.38),

$$\left(\frac{D_f}{B}\right)_{cr-R} = \left(\frac{D_f}{B}\right)_{cr-S}\left[0.73 + 0.27\left(\frac{L}{B}\right)\right] = (7)\left[0.73 + 0.27\left(\frac{3}{1.5}\right)\right] = 8.89$$

Check:

$$1.55\left(\frac{D_f}{B}\right)_{cr\text{-}S} = (1.55)(7) = 10.85$$

So, use $(D_f/B)_{cr\text{-}R} = 8.89$. The actual embedment ratio is $D_f/B = 1.8/1.5 = 1.2$. Hence, this is a shallow foundation:

$$\alpha' = \frac{\dfrac{D_f}{B}}{\left(\dfrac{D_f}{B}\right)_{cr}} = \frac{1.2}{8.89} = 0.13$$

Referring to the average curve of Figure 8.20 for $\alpha' = 0.13$, the magnitude of $\beta' = 0.2$. From equation (8.43),

$$Q_u = A\left\{\beta'\left[7.56 + 1.44\left(\frac{B}{L}\right)\right]c_u + \gamma D_f\right\} = (1.5)(3)\left\{(0.2)\left[7.56 + 1.44\left(\frac{1.5}{3}\right)\right](52) + (18.9)(1.8)\right\}$$

$$= 540.6\ kN$$

8.3.5 THREE-DIMENSIONAL LOWER BOUND SOLUTION

Merifield et al.[19] used a three-dimensional numerical procedure based on a finite element formulation of the lower bound theorem of limit analysis to estimate the uplift capacity of foundations. The results of this study, along with the procedure to determine the uplift capacity, are summarized below in a step-by-step manner.

1. Determine the breakout factor in a homogeneous soil with no unit weight (that is, $\gamma = 0$) as

$$F_c = F_{co} \tag{8.44}$$

   The variation of F_{co} for square, circular, and rectangular foundations is shown in Figure 8.21.

2. Determine the breakout factor in a homogeneous soil with unit weight (that is, $\gamma \neq 0$) as

$$F_c = F_{c\gamma} = F_{co} + \frac{\gamma D_f}{c_u} \tag{8.45}$$

3. Determine the breakout factor for a deep foundation $F_c = F_c^*$ as follows:

   $F_c^* = 12.56$ (for circular foundations)
   $F_c^* = 11.9$ (for square foundations)
   $F_c^* = 11.19$ (for strip foundations with $L/B \geq 10$)

4. If $F_{c\gamma} \geq F_c^*$, it is a deep foundation. Calculate the ultimate load as

$$Q_u = Ac_u F_c^* \tag{8.46}$$

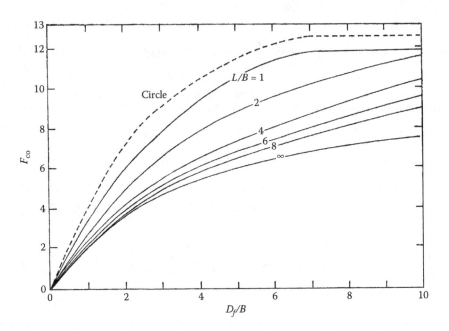

FIGURE 8.21 Numerical lower bound solution of Merifield et al.—plot of F_{co} versus D_f/B for circular, square, and rectangular foundations. *Source:* Merifield, R. S., A. V. Lyamin, S. W. Sloan, and H. S. Yu. 2003. Three-dimensional lower bound solutions for stability of plate anchors in clay. *J. Geotech. Geoenv. Eng.*, ASCE, 129(3): 243.

However, if $F_{c\gamma} \leq F_c^*$, it is a shallow foundation. Thus,

$$Q_u = Ac_u F_{c\gamma} \tag{8.47}$$

EXAMPLE 8.5

Solve the Example 8.4 problem using the procedure outlined in section 8.35.

Solution

Given: $L/B = 3/1.5 = 2$; $D_f/B = 1.8/1.5 = 1.2$. From Figure 8.21, for $L/B = 2$ and $D_f/B = 1.2$, the value of $F_{co} \approx 3.1$:

$$F_{c\gamma} = F_{co} + \frac{\gamma D_f}{c_u} = 3.1 + \frac{(18.9)(1.8)}{52} = 3.754$$

For a foundation with $L/B = 2$, the magnitude of $F_c^* \approx 11.5$. Thus,

$$F_{c\gamma} < F_c^*$$

Hence,

$$Q_u = Ac_u F_{c\gamma} = (3 \times 1.5)(52)(3.754) \approx \mathbf{878 \ kN}$$

8.3.6 FACTOR OF SAFETY

In most cases of foundation design, it is recommended that a minimum factor of safety of 2 to 2.5 be used to arrive at the allowable ultimate uplift capacity.

REFERENCES

1. Balla, A. 1961. The resistance to breaking out of mushroom foundations for pylons, in *Proc., V Int. Conf. Soil Mech. Found. Eng.*, Paris, France, 1: 569.
2. Meyerhof, G. G., and J. I. Adams. 1968. The ultimate uplift capacity of foundations. *Canadian Geotech. J.* 5(4): 225.
3. Caquot, A., and J. Kerisel. 1949. *Tables for calculation of passive pressure, active pressure, and bearing capacity of foundations.* Paris: Gauthier-Villars.
4. Das, B. M., and G. R. Seeley. 1975. Breakout resistance of horizontal anchors. *J. Geotech. Eng. Div.*, ASCE, 101(9): 999.
5. Meyerhof, G. G. 1973. Uplift resistance of inclined anchors and piles, in *Proc., VIII Int. Conf. Soil Mech. Found. Eng.*, Moscow, USSR, 2.1: 167.
6. Das, B. M., and A. D. Jones. 1982. Uplift capacity of rectangular foundations in sand. *Trans. Res. Rec. 884*, National Research Council, Washington, D.C. 54.
7. Vesic, A. S. 1965. Cratering by explosives as an earth pressure problem, in *Proc., VI Int. Conf. Soil Mech. Found. Eng.*, Montreal, Canada, 2: 427.
8. Vesic, A. S. 1971. Breakout resistance of objects embedded in ocean bottom. *J. Soil Mech. Found. Div.*, ASCE 97(9): 1183.
9. Saeedy, H. S. 1987. Stability of circular vertical earth anchors. *Canadian Geotech. J.* 24(3): 452.
10. Baker, W. H., and R. L. Kondner. 1966. Pullout load capacity of a circular earth anchor buried in sand. *Highway Res. Rec.108*, National Research Council, Washington, D.C. 1.
11. Sutherland, H. B. 1965. Model studies for shaft raising through cohesionless soils, in *Proc., VI Int. Conf. Soil Mech. Found. Eng.*, Montreal Canada, 2: 410.
12. Esquivel-Diaz, R. F. 1967. Pullout resistance of deeply buried anchors in sand. M.S. Thesis, Duke University, Durham, NC, USA.
13. Das, B. M. 1978. Model tests for uplift capacity of foundations in clay. *Soils and Foundations*, Japan 18(2): 17.
14. Ali, M. 1968. Pullout resistance of anchor plates in soft bentonite clay. M.S. Thesis, Duke University, Durham, NC, USA.
15. Kupferman, M. 1971. The vertical holding capacity of marine anchors in clay subjected to static and dynamic loading, M.S. Thesis, University of Massachusetts, Amherst, MA, USA.
16. Adams, J. K., and D. C. Hayes. 1967. The uplift capacity of shallow foundations. *Ontario Hydro. Res. Quarterly* 19(1): 1.
17. Bhatnagar, R. S. 1969. Pullout resistance of anchors in silty clay. M.S. Thesis, Duke University, Durham, NC, USA.
18. Das, B. M. 1980. A procedure for estimation of ultimate uplift capacity of foundations in clay. *Soils and Foundations*, Japan, 20(1): 77.
19. Merifield, R. S., A. V. Lyamin, S. W. Sloan, and H. S. Yu. 2003. Three-dimensional lower bound solutions for stability of plate anchors in clay. *J. Geotech. Geoenv. Eng.*, ASCE, 129(3): 243.

Index

Printed in the United States
by Baker & Taylor Publisher Services